Physiology at a Glance

KU-620-365

WITHDRAWN

LIVERPOOL JMU LIBRARY

3 1111 01435 1637

This new edition is also available as an e-book.
For more details, please see
www.wiley.com/buy/9780470659786
or scan this QR code:

Physiology at a Glance

Jeremy P.T. Ward

PhD
Head of Department of Physiology
and Professor of Respiratory Cell Physiology
Schools of Medicine and Biomedical Sciences
King's College London
London

Roger W.A. Linden

BDS PhD MFDS RCS
Emeritus Professor of Craniofacial Biology
School of Biomedical Sciences
King's College London
London

Third Edition

A John Wiley & Sons, Ltd., Publication

This edition first published 2013 © 2013 by John Wiley & Sons, Ltd

Wiley-Blackwell is an imprint of John Wiley & Sons, formed by the merger of Wiley's global Scientific, Technical and Medical business with Blackwell Publishing.

Registered office: John Wiley & Sons, Ltd, The Atrium, Southern Gate, Chichester, West Sussex, PO19 8SQ, UK

Editorial offices: 9600 Garsington Road, Oxford, OX4 2DQ, UK
The Atrium, Southern Gate, Chichester, West Sussex, PO19 8SQ, UK
111 River Street, Hoboken, NJ 07030-5774, USA

For details of our global editorial offices, for customer services and for information about how to apply for permission to reuse the copyright material in this book please see our website at www.wiley.com/wiley-blackwell.

The right of the author to be identified as the author of this work has been asserted in accordance with the UK Copyright, Designs and Patents Act 1988.

All rights reserved. No part of this publication may be reproduced, stored in a retrieval system, or transmitted, in any form or by any means, electronic, mechanical, photocopying, recording or otherwise, except as permitted by the UK Copyright, Designs and Patents Act 1988, without the prior permission of the publisher.

Designations used by companies to distinguish their products are often claimed as trademarks. All brand names and product names used in this book are trade names, service marks, trademarks or registered trademarks of their respective owners. The publisher is not associated with any product or vendor mentioned in this book.

Limit of Liability/Disclaimer of Warranty: While the publisher and author(s) have used their best efforts in preparing this book, they make no representations or warranties with respect to the accuracy or completeness of the contents of this book and specifically disclaim any implied warranties of merchantability or fitness for a particular purpose. It is sold on the understanding that the publisher is not engaged in rendering professional services and neither the publisher nor the author shall be liable for damages arising herefrom. If professional advice or other expert assistance is required, the services of a competent professional should be sought.

Library of Congress Cataloging-in-Publication Data

Ward, Jeremy P. T.
  Physiology at a glance / Jeremy P.T. Ward, Roger W.A. Linden. – 3rd ed.
       p. ; cm. – (At a glance)
  Includes bibliographical references and index.
  ISBN 978-0-470-65978-6 (pbk. : alk. paper)
  I. Linden, R. W. A. (Roger W. A.) II. Title. III. Series: At a glance series (Oxford, England)
  [DNLM: 1. Physiological Phenomena–Handbooks.   QT 29]
  612–dc23
                            2012044842

A catalogue record for this book is available from the British Library.

Wiley also publishes its books in a variety of electronic formats. Some content that appears in print may not be available in electronic books.

Cover image: iStockphoto.com/Eraxion
Cover design by Meaden Creative

Set in 9/11.5 pt Times by Toppan Best-set Premedia Limited
Printed and bound in Malaysia by Vivar Printing Sdn Bhd

1   2013

Contents

Preface

Physiology is defined as 'the scientific study of the bodily function of living organisms and their parts'. There is a natural symbiosis between function (physiology) and structure (anatomy) from which physiology emerged as a separate discipline in the late 19th century. A good understanding of anatomy and physiology is an essential pre-requisite for understanding what happens when things go wrong – the structural abnormalities and pathophysiology of disease – and as such underpins all biomedical studies and medicine itself. Following a century of reductionism, where the focus of research has progressively narrowed down to the function of individual proteins and genes, there is now a resurgence in integrative physiology, as it has been realized that to make sense of developments such as the Human Genome Project we have to understand body function as an integrated whole. This is considerably more complex than just the sum of its parts because of the multiplicity of interactions involved. True understanding of the role of a single gene, for example, can only be gained when placed in the context of the whole animal, as reflected by the often unexpected effects of knock-out of single genes on the phenotype of mice.

This volume is designed as a concise guide and revision aid to core topics in physiology, and should be useful to all students following a first-year physiology course, whether they are studying single honours, biomedical sciences, nursing, medicine or dentistry. It should also be useful to those studying system-based curricula. The layout of *Physiology at a Glance* follows that of the other volumes in the *At a Glance* series, with a two-page spread for each topic (loosely corresponding to a lecture), comprising a large diagram on one page and concise explanatory text on the other. For this third edition we have exten-sively revised the text and figures, and some chapters have been completely replaced. The chapter on Blood has been split into two, so that more space could be given to erythropoiesis and haemostasis, important but sometimes difficult to understand mechanisms. We have also corrected some embarrassing errors.

Physiology is a large subject, and in a book this size we cannot hope to cover anything but the core and basics. *Physiology at a Glance* should therefore be used primarily to assist basic understanding of key concepts and as an assistance to revision. Deeper knowledge should be gained by reference to full physiology and system textbooks, and in third-year honours programmes to original peer-reviewed papers. Students may find one or two sections of this book difficult, such as that on the physics of flow and diffusion, and such material may indeed not be included in some introductory physiology courses. However, understanding these concepts often assists in learning how body systems behave in the way they do.

In revising this third edition we have been helped immensely by constructive criticism and suggestions from our colleagues and students, and junior and senior reviewers of the last edition. We would particularly like to mention Professor Peter Jones and Dr James Bowe for kindly assisting us with revision of the endocrine and reproduction sections, and Dr Liz Andrew for assistance with the immunology. We thank all those who have given us such advice; any errors are ours and not theirs. We would also like to thank the team at Wiley-Blackwell who provided great encouragement and support throughout the project.

Jeremy Ward
Roger Linden

Acknowledgements

Some figures in this book are taken from:

Ward, J.P.T., Ward, J., and Leach, R.M. (2010) *The Respiratory System at a Glance (3rd edition)*. Wiley-Blackwell, Oxford.

Aaronson, P.I., Ward, J.P.T. and Connolly, M.J. (2012) *The Cardiovascular System at a Glance (4th edition)*. Wiley-Blackwell, Oxford.

Mehta, A. and Hoffbrand, V. (2009) *Haematology at a Glance (3rd edition)*. Wiley-Blackwell, Oxford.

List of abbreviations

1,25-(OH)$_2$D	1,25-dihydroxycholecalciferol
2,3-DPG	2,3-diphosphoglycerate
5-HT	5-hydroxytryptamine; serotonin
ACE	angiotensin-converting enzyme
ACh	acetylcholine
ACTH	adrenocorticotrophic hormone
ADH	antidiuretic hormone
ADP	adenosine diphosphate
AIDS	acquired immune deficiency syndrome
ANP	atrial natriuretic peptide
ANS	autonomic nervous system
AP	action potential
ATP	adenosine triphosphate
ATPase	enzyme that splits ATP
AV node	atrioventricular node (heart)
BTPS	body temperature and pressure, saturated with water
CAM-kinase	calcium–calmodulin kinase
cAMP	cyclic adenosine monophosphate
CCK	cholecystokinin
CDI	central diabetes insipidus
cGMP	cyclic guanosine monophosphate
CICR	Ca^{2+}-induced Ca^{2+} release
COMT	catechol-O-methyl transferase
COX	cyclooxygenase
CRH	corticotrophin-releasing hormone
CVP	central venous pressure
Da	Dalton (unit for molecular weight)
DAG	diacylglycerol
DHEA	dehydroepiandrosterone
D$_{LO_2}$	O_2 diffusing capacity in lung; transfer factor
DNA	deoxyribonucleic acid
DOPA	dihydroxyphenylalanine
E$_{(ion)}$	equilibrium potential for ion (K^+, Na^+, Ca^{2+} or Cl^-)
ECF	extracellular fluid
ECG (EKG)	electrocardiogram (or graph)
EDP	end diastolic pressure
EDV	end diastolic volume
EGF	epidermal growth factor
E$_m$	membrane potential
EPO	erythropoietin
EPP	end plate potential
ESV	end systolic volume
F$_{ab}$	hypervariable region of antibody molecule
F$_c$	constant region of antibody molecule
FEV$_1$	forced expiratory volume in 1 s
FGF	fibroblast growth factor
F$_{N_2}$	fractional concentration of nitrogen in a gas mixture
FSH	follicle-stimulating hormone
FVC	forced vital capacity
GDP	guanosine diphosphate
GFR	glomerular filtration rate
GH	growth hormone
GHRH	growth hormone-releasing hormone
GIP	gastric inhibitory peptide
GLUT-1, 2 or 4	glucose transporters
GnRH	gonadotrophin-releasing hormone
GPCR	G-protein-coupled receptor
G-protein	GTP-binding protein
GRP	gastrin-releasing peptide
GTP	guanosine triphosphate
GTPase	enzyme that splits GTP
hCG	human chorionic gonadotrophin
HIV	human immunodeficiency virus
HMWK	high molecular weight kininogen
ICF	intracellular fluid
IgA, E, G, M	immunoglobulin A, E, G or M
IGF-1 or 2	insulin-like growth factor (1 or 2)
IL-1b or 6	interleukin-1β or 6
IP$_3$	inositol triphosphate
IRS-1	insulin receptor substrate 1
ISF	interstitial fluid
JAK	Janus kinase
JGA	juxtaglomerular apparatus
LH	luteinizing hormone
MAO	monoamine oxidase
MAP	mean arterial pressure
MAPK(K)	mitogen-activated protein kinase (kinase)
MEPP	miniature end plate potentials
MIH	melanotrophin-inhibiting hormone
mRNA	messenger RNA
MSH	melanotrophin-stimulating hormone
Na$^+$ pump	Na^+-K^+ ATPase
NAD$^+$ or (H)	nicotinic adenine dinucleotide (oxidized and reduced forms)
NDI	nephrogenic diabetes insipidus
NGF	nerve growth factor
NO	nitric oxide
NOS	nitric oxide synthase
PAH	*para*-aminohippuric acid
PDGF	platelet-derived growth factor
PI-3 kinase	phosphatidylinositol-3 kinase
pK	negative log of dissociation constant (buffers)
PKC	protein kinase C
PTH	parathyroid hormone
Ras	a cellular GTPase
ROC	receptor-operated channels
RTK	receptor tyrosine kinase
SA node	sinoatrial node
SERCA	smooth endoplasmic reticulum Ca^{2+} ATPase
SMAD	intracellular protein associated with streptokinases
SOC	store-operated channels
SR	sarcoplasmic reticulum
SST	somatostatin
STAT	signal transduction and activation of transcription (protein)
S-TK	serine–threonine kinase
STPD	standard temperature and pressure, dry gas

$T_{1 \text{ or } 2}$	mono- or di-iodotyrosine	TRE	thyroid response element
T_3	tri-iodothyronine	tRNA	transfer RNA
T_4	thyroxine	TSH	thyroid-stimulating hormone
TGFβ	transforming growth factor β	TV	tidal volume
TH	thyroid hormone	TXA_2	thromboxane A_2
T_m	tubular transport maximum (kidney)	UCP-1, 2 or 3	uncoupling protein-1, 2 or 3
TNF	tumour necrosis factor	V_A/Q mismatch	ventilation–perfusion mismatch (lungs)
TRa	thyroid hormone receptor	VIP	vasoactive intestinal polypeptide

About the companion website

There is a companion website for this book at:

www.ataglanceseries.com/physiology

The website will feature:
– MCQs for each chapter
– Interactive self-test flashcards
– Revision notes

1 Homeostasis and the physiology of proteins

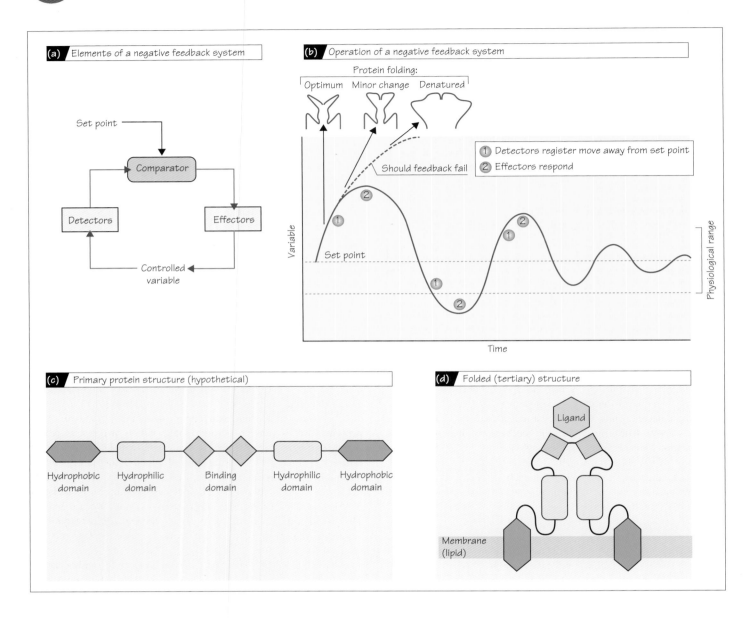

(a) Elements of a negative feedback system

Set point

Comparator

Detectors

Effectors

Controlled variable

(b) Operation of a negative feedback system

Protein folding:

Optimum   Minor change   Denatured

① Detectors register move away from set point
② Effectors respond

Should feedback fail

Variable

Set point

Physiological range

Time

(c) Primary protein structure (hypothetical)

Hydrophobic domain   Hydrophilic domain   Binding domain   Hydrophilic domain   Hydrophobic domain

(d) Folded (tertiary) structure

Ligand

Membrane (lipid)

Claude Bernard (1813–1878) first described 'le mileau intérieur' and observed that the internal environment of the body remained remarkably constant (or in equilibrium) despite the ever changing external environment. The term **homeostasis** was not used until 1929 when **Walter Cannon** first used it to describe this ability of physiological systems to maintain conditions within the body in a relatively constant state of equilibrium. It is arguably the most important concept in physiology.

Homeostasis is Greek for 'staying the same'. However, this so-called **equilibrium** is not an unchanging state but is a dynamic state of equilibrium causing a **dynamic constancy** of the internal environment. This **dynamic constancy** arises from the variable responses caused by changes in the external environment. Homeostasis maintains most physiological systems and examples are seen throughout this book. The way in which the body maintains the H^+ ion concentration of body fluids within narrow limits, the control of blood glucose

Physiology at a Glance, Third Edition. Jeremy P.T. Ward and Roger W.A. Linden.

 © 2013 John Wiley & Sons, Ltd. Published 2013 by John Wiley & Sons, Ltd.

by the release of insulin, and the control of body temperature, heart rate and blood pressure are all examples of homeostasis. The human body has literally thousands of control systems. The most intricate are genetic control systems that operate in all cells to control intracellular function as well as all extracellular functions. Many others operate within organs to control their function; others operate throughout the body to control interaction between organs. As long as conditions are maintained within the normal physiological range within the internal environment, the cells of the body continue to live and function properly. Each cell benefits from homeostasis and in turn, each cell contributes its share towards the maintenance of homeostasis. This reciprocal interplay provides continuity of life until one or more functional systems lose their ability to contribute their share. Moderate dysfunction of homeostasis leads to sickness and disease, and extreme dysfunction of homeostasis leads to death.

Negative feedback control

Most physiological control mechanisms have a common basic structure. The factor that is being controlled is called the **variable.** Homeostatic mechanisms provide the tight regulation of *all* physiological variables and the most common type of regulation is by *negative* **feedback**. A negative feedback system (Fig. 1a) comprises: **detectors** (often neural **receptor cells**) to measure the variable in question; a **comparator** (usually a neural assembly in the central nervous system) to receive input from the detectors and compare the size of the signal against the desired level of the variable (the **set point**); and **effectors** (muscular and/or glandular tissue) that are activated by the comparator to restore the variable to its set point. The term 'negative feedback' comes from the fact that the effectors always act to move the variable in the opposite direction to the change that was originally detected. Thus, when the partial pressure of CO_2 in blood increases above 40 mmHg, brain stem mechanisms increase the rate of ventilation to clear the excess gas, and vice versa when CO_2 levels fall below 40 mmHg (Chapter 29). The term 'set point' implies that there is a single optimum value for each physiological variable; however, there is some tolerance in all physiological systems and the set point is actually a narrow *range* of values within which physiological processes will work normally (Fig. 1b). Not only is the set point not a point, but it can be reset in some systems according to physiological requirements. For instance, at high altitude, the low partial pressure of O_2 in inspired air causes the ventilation rate to increase. Initially, this effect is limited due to the loss of CO_2, but, after 2–3 days, the brain stem lowers the set point for CO_2 control and allows ventilation to increase further, a process known as **acclimatization**.

A common operational feature of all negative feedback systems is that they induce oscillations in the variable that they control (Fig. 1b). The reason for this is that it takes time for a system to detect and respond to a change in a variable. This delay means that feedback control always causes the variable to overshoot the set point slightly, activating the opposite restorative mechanism to induce a smaller overshoot in that direction, until the oscillations fall within the range of values that are optimal for physiological function. Normally, such oscillations have little visible effect. However, if unusually long delays are introduced into a system, the oscillations can become extreme. Patients with congestive heart failure sometimes show a condition known as **Cheyne–Stokes' breathing**, in which the patient undergoes periods of deep breathing interspersed with periods of no breathing at all (**apnoea**). This is partly due to the slow flow of blood from the lungs to the brain, which causes a large delay in the detection of blood levels of CO_2.

Some physiological responses use *positive* **feedback**, causing rapid amplification. Examples include initiation of an action potential, where sodium entry causes depolarization which further increases sodium entry and thus more depolarization (Chapter 5), and certain hormonal changes, particularly in reproduction (Chapter 50). Positive feedback is inherently unstable, and requires some mechanism to break the feedback loop and stop the process, such as time-dependent inactivation of sodium channels in the first example and the birth of the child in the second.

Protein form and function are protected by homeostatic mechanisms

The homeostatic mechanisms that are described in detail throughout this book have evolved to protect the integrity of the protein products of gene translation. Normal functioning of proteins is essential for life, and usually requires binding to other molecules, including other proteins. The specificity of this binding is determined by the three-dimensional shape of the protein. The **primary structure** of a protein is determined by the sequence of amino acids (Fig. 1c). Genetic mutations that alter this sequence can have profound effects on the functionality of the final molecule. Such gene **polymorphisms** are the basis of many genetically based disorders. The final shape of the molecule (the **tertiary structure**), however, results from a process of **folding** of the amino acid chain (Fig. 1d). Folding is a complex process by which a protein achieves its lowest energy conformation. It is determined by electrochemical interactions between amino acid side-chains (e.g. hydrogen bonds, van der Waals' forces), and is so vital that it is overseen by **molecular chaperones**, such as the **heat shock proteins**, which provide a quiet space within which the protein acquires its final shape. In healthy tissue, cells can detect and destroy misfolded proteins, the accumulation of which damages cells and is responsible for various pathological conditions, including **Alzheimer's disease** and **Creutzfeldt–Jakob disease**. Folding ensures that the functional sequences of amino acids (**domains**) that form, e.g. binding sites for other molecules or hydrophobic segments for insertion into a membrane, are properly orientated to allow the protein to serve its function.

The relatively weak nature of the forces that cause folding renders them sensitive to changes in the environment surrounding the protein. Thus, alterations in acidity, osmotic potential, concentrations of specific molecules/ions, temperature or even hydrostatic pressure can modify the tertiary shape of a protein and change its interactions with other molecules. These modifications are usually reversible and are exploited by some proteins to detect alterations in the internal or external environments. For instance, nerve cells that respond to changes in CO_2 (chemoreceptors; Chapter 29) possess **ion channel** proteins (Chapter 4) that open or close to generate electrical signals (Chapter 5) when the acidity of the medium surrounding the receptor (CO_2 forms an acid in solution) alters by more than a certain amount. However, there are limits to the degree of fluctuation in the internal environment that can be tolerated by proteins before their shape alters so much that they become non-functional or irreversibly damaged, a process known as **denaturation** (this is what happens to egg-white proteins in cooking). Homeostatic systems prevent such conditions from arising within the body, and thus preserve protein functionality.

2 Body water compartments and physiological fluids

(a) Physiological fluid compartments

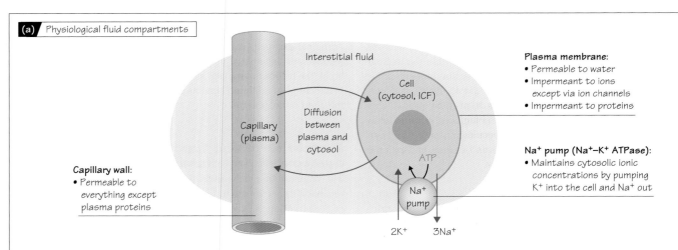

Constituents of physiological fluids (approximate values, intracellular varies between tissues)		Plasma	Interstitial	Intracellular	Unit
Water:	% total body water	13%	22%	65%	%
	(volume in a 70-kg person)	(3.5)	(9.5)	(27)	L
Osmolality		290	290	290	mosmol/kg H_2O
Cations:	Na^+	140	140	10	mmol/L
	K^+	4	4	140	mmol/L
	Ca^{2+} (free)	1	1	0.0001	mmol/L
Anions:	Cl^-	108	129	3–30	mmol/L
	HCO_3^-	26	26	9	mmol/L
	Proteins$^-$	10	1	50	mmol/L
	Other anions (mainly PO_4^{3-}, SO_4^{3-})	3	0	60–88	mmol/L

Notes: Ca^{2+} (and Mg^{2+}) tend to bind to plasma proteins, and their free concentrations are about 50% of the total. Ionic concentrations are sometimes given in mEq/L to reflect the amount of charge, where an equivalent (Eq) is 1 mole of charge. So 1 Eq of a monovalent ion such as Na^+ = 1 mole, but 1 Eq of Ca^{2+} = 0.5 mole

(b) Effects of ingesting fluids of differing osmotic potential

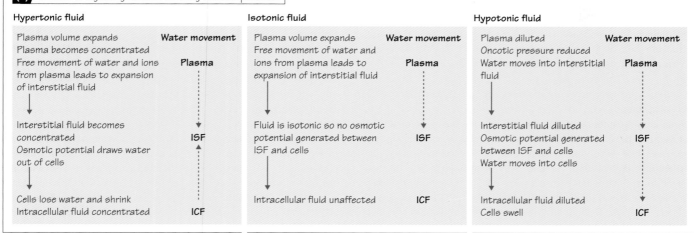

Hypertonic fluid

Plasma volume expands
Plasma becomes concentrated
Free movement of water and ions from plasma leads to expansion of interstitial fluid

↓

Interstitial fluid becomes concentrated
Osmotic potential draws water out of cells

↓

Cells lose water and shrink
Intracellular fluid concentrated

Water movement

Plasma

ISF

ICF

Isotonic fluid

Plasma volume expands
Free movement of water and ions from plasma leads to expansion of interstitial fluid

↓

Fluid is isotonic so no osmotic potential generated between ISF and cells

↓

Intracellular fluid unaffected

Water movement

Plasma

ISF

ICF

Hypotonic fluid

Plasma diluted
Oncotic pressure reduced
Water moves into interstitial fluid

↓

Interstitial fluid diluted
Osmotic potential generated between ISF and cells
Water moves into cells

↓

Intracellular fluid diluted
Cells swell

Water movement

Plasma

ISF

ICF

Physiology at a Glance, Third Edition. Jeremy P.T. Ward and Roger W.A. Linden.

© 2013 John Wiley & Sons, Ltd. Published 2013 by John Wiley & Sons, Ltd.

Osmosis

Osmosis is the passive movement of water across a **semi-permeable membrane** from regions of low solute concentration to those of higher solute concentration. Biological membranes are semi-permeable in that they usually allow the free movement of water but restrict the movement of solutes. The creation of **osmotic gradients** is the primary method for the movement of water in biological systems. This is why the osmotic potential (**osmolality**) of body fluids is closely regulated by a number of homeostatic mechanisms (Chapter 35). A fluid at the same osmotic potential as plasma is said to be **isotonic**; one at higher potential (i.e. more concentrated solutes) is **hypertonic** and one at lower potential is **hypotonic**. The osmotic potential depends on the number of osmotically active particles (molecules) per litre, irrespective of their identity. It is expressed in terms of osmoles, where 1 osmole equals 1 mole of particles, as **osmolarity** (osmol/L), or **osmolality** (osmol/kg H_2O). The latter is preferred by physiologists as it is independent of temperature, though in physiological fluids the values are very similar. The osmolality of plasma is ~290 mosmol/kg H_2O, mostly due to dissolved ions and small molecules (e.g. glucose and urea). These diffuse easily across capillaries, and the **crystalloid osmotic pressure** they exert is therefore the same either side of the capillary wall. Proteins do not easily pass through capillary walls, and are responsible for the **oncotic** (or colloidal osmotic) pressure. This is much smaller than crystalloid osmotic pressure, but is critical for fluid transfer across capillary walls because it differs between plasma and interstitial fluid (Chapter 23). Oncotic pressure is expressed in terms of pressure, and in plasma is normally ~25 mmHg. Maintenance of plasma osmolality is vital for regulation of blood volume (Chapter 22). Drinking fluids of differing osmotic potentials has distinct effects on the distribution of water between cells and extracellular fluid (Fig. 2b).

Body water compartments

Water is the solvent in which almost all biological reactions take place (the other being membrane lipid), and so it is fitting that it accounts for some 50–70% of the body mass (i.e. about 40 L in a 70-kg person). The nature of biological membranes means that water moves freely within the body, but the materials dissolved in it do not. There are two major 'fluid compartments': the water within cells (**intracellular fluid, ICF**), which accounts for about 65% of the body total, and the water outside cells (**extracellular fluid, ECF**). These compartments are separated by the plasma membranes of the cells, and differ markedly in terms of the concentrations of the ions that are dissolved in them (Fig. 2a; Chapter 4). Approximately 65% of the ECF comprises the tissue fluid found between cells (**interstitial fluid, ISF**), and the rest is made up of the liquid component of blood (**plasma**). The barrier between these two fluids consists of the walls of the capillaries (Fig. 2a; Chapter 23).

Intracellular versus extracellular fluid

Many critical biological events, including all bioelectrical signals (Chapter 5), depend on maintaining the composition of physiological fluids within narrow limits. Figure 2a shows the concentrations of ions in the three main fluid compartments. It should be noted that, *within* any one compartment, there *must* be electrical neutrality, i.e. the total number of positive charges must equal the total number of negative charges. The most important difference between ICF and ECF lies in the relative concentrations of cations. The K^+ ion concentration is much higher inside the cell than in ECF, while the opposite is true for the Na^+ ion concentration. Ca^{2+} and Cl^- ion concentrations are also higher in ECF. The question arises as to how these differences come about, and how they are maintained. Ion channel proteins allow the cell to determine the flow of ions across its own membrane (Chapter 4). In most circumstances, relatively few channels are open so that the leakage of ions is low. There is, however, always a steady movement of ions across the membrane, with Na^+ and K^+ following their concentration gradients into and out of the cell, respectively. Uncorrected, the leak would eventually lead to the equalization of the compositions of the two compartments, effectively eliminating all bioelectrical signalling (Chapter 5). This is prevented by the activity of the Na^+-K^+ ATPase, or Na^+ pump (Chapter 3). Of the other ions, most Ca^{2+} in the cell is transported actively either out of the cell or into the endoplasmic reticulum and mitochondria, leaving very low levels of free Ca^{2+} in ICF. Cl^- ions are differentially distributed across the membrane by virtue of their negative charge. Intracellular proteins are negatively charged at physiological pH. These and other large anions that cannot cross the plasma membrane (e.g. phosphate, PO_4^{3-}) are trapped within the cell and account for most of the anion content of ICF. Cl^- ions, which *can* diffuse across the membrane through channels, are forced out of the cell by the charge on the fixed anions. The electrical force driving Cl^- ions out of the cell is balanced by the chemical gradient driving them back in, a situation known as the **Gibbs–Donnan equilibrium**. Variations in the large anion content of cells mean that the concentration of Cl^- ions in ICF can vary by a factor of 10 between cell types, being as high as 30 mM in cardiac myocytes, although lower values (around 5 mM) are more common.

Interstitial fluid versus plasma

The main difference between these fluids is that plasma contains more protein than does ISF (Fig. 2a). The plasma proteins (Chapter 8) are the only constituents of plasma that do not cross into ISF, although they are allowed to escape from capillaries in very specific circumstances (Chapter 10). The presence of impermeant proteins in the plasma exerts an osmotic force relative to ISF (**plasma oncotic pressure**; see above) that almost balances the hydrostatic pressure imposed on the plasma by the action of the heart, which tends to force water out of the capillaries, so that there is a small net water movement out of the plasma into the interstitial space. The leakage is absorbed by the **lymphatic system** (Chapter 23). **Transcellular fluid** is the name given to fluids that do not contribute to any of the main compartments, but which are derived from them. It includes cerebrospinal fluid and exocrine secretions, particularly gastrointestinal secretions (Chapters 37–41), and has a collective volume of approximately 2 L.

VERPOOL JOHN MOORES UNIVERSITY
LEARNING SERVICES

3 Cells, membranes and organelles

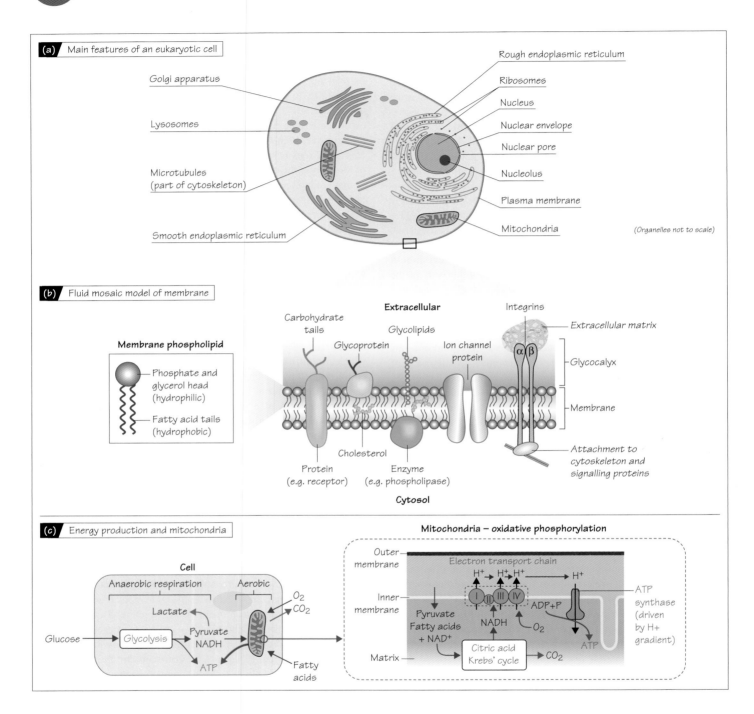

Physiology at a Glance, Third Edition. Jeremy P.T. Ward and Roger W.A. Linden.
16 © 2013 John Wiley & Sons, Ltd. Published 2013 by John Wiley & Sons, Ltd.

The aqueous internal environment of the cell is separated from the aqueous external medium by an envelope of fat molecules (**lipids**) known as the **plasma membrane**. About half the cell is filled with **cytosol**, a viscous, protein-rich fluid between the internal structures. These consist of **organelles** which are themselves enclosed by lipid membranes, and components of the **cytoskeleton** such as microtubules and actin filaments which provide structural stability. The reticular appearance of the cell interior is due to organelles whose membranes are folded to maximize surface area. These include the **rough endoplasmic reticulum** and **Golgi apparatus**, which are involved in protein assembly, and the **smooth endoplasmic reticulum** which serves as a store for intracellular Ca^{2+} and is the major site of lipid production (Fig. 3a).

Protein-processing organelles

The **nucleus** (Fig. 3a) contains the chromosomes and **nucleolus**, a membrane-less structure responsible for production of **ribosomes**. Ribosomes translocate to the **rough endoplasmic reticulum** (giving it its appearance), where they are responsible for protein assembly. The endoplasmic reticulum and **Golgi apparatus** perform post-translational processing of new proteins. This includes trimming amino acid chains to the right length, protein folding, addition of polysaccharide chains (**glycosylation**) and identification of improperly folded proteins, which are tagged for subsequent destruction by lysosomes. Proteins are delivered from the Golgi apparatus to specific intracellular destinations. For example, receptor and structural proteins are sent to the membrane and digestive enzymes to lysosomes, and molecules for extracellular action are packaged into secretory vesicles. **Lysosomes** contain acid hydrolase enzymes which catabolize macromolecules. They work optimally at pH 5.0, and as cytosolic pH is ~7.2, any leaking into the cytosol cannot attack the cell inappropriately. Lysosomes digest unwanted and defective proteins, recycling raw materials and preventing accumulation of rubbish.

Membranes and membrane proteins

Membrane lipids (mostly phospholipids) comprise a **hydrophilic** (water-loving) head, with two short **hydrophobic** (water-repelling) fatty acid tails (Fig. 3b). In an aqueous medium they self-organize into a **bilayer** with the heads facing outwards and the tails inwards (Fig. 3b). They diffuse freely within each layer (**lateral diffusion**) so the membrane is fluid. The hydrophobic interior and hydrophilic exterior of the membrane means that lipid-soluble (hydrophobic) substances such as **cholesterol** incorporate into the membrane, whilst molecules with both hydrophobic and hydrophilic domains such as proteins can be tethered part in and part out of the membrane (the **fluid mosaic model**; Fig 3b). Many such molecules provide signalling, transport or structural functions. The latter are provided by proteins such as **spectrin**, which binds to the inner layer and forms an attachment framework for the **cytoskeleton**. Lipid-soluble molecules such as O_2 and CO_2, and small molecules such as water and urea readily pass through the lipid bilayer. However, larger molecules such as glucose and polar (charged) molecules such as ions cannot, and their transport is mediated by **transporter** and **ion channel** membrane proteins (Chapter 4). Membrane proteins also undergo lateral diffusion and move around the membrane. However, the cell can control exactly which proteins insert into which portion of the membrane. For example, cells lining the kidney tubules are polarized so that $Na^+ - K^+$ **ATPase** transporters (Chapters 4 and 33) are located only on one side of the cell. Most cells are covered by a thin gel-like layer called the **glycocalyx**, containing glycoproteins and carbohydrate chains extending from the membrane and secreted proteins (Fig. 3b). It protects the membrane and also plays a role in cell function and cell–cell interactions.

Membrane proteins associated with cell signalling include enzymes bound to the inner surface such as **phospholipases**, which produce arachidonic acid (a precursor of some second messenger molecules), and **adenylyl cyclase**, which generates the second messenger cyclic adenosine monophosphate (**cAMP**). cAMP activates **protein kinase** enzymes to initiate numerous changes in cell function by **phosphorylating** membrane and intracellular proteins. **Transmembrane** proteins (Fig. 3b) penetrate the entire thickness of the bilayer, and include **receptors** and **ion channel proteins**. The intramembrane segments are composed of hydrophobic amino acid residues and the extra- and intra-cellular portions predominantly of hydrophilic residues. Receptors include those that bind **growth factors** and regulate gene transcription, and the superfamily known as **G-protein–coupled receptors** (**GPCRs**). The latter possess seven membrane-spanning segments and detect neurotransmitters or hormones in the extracellular medium. On binding the appropriate molecule, they activate specific membrane-associated **GTP-binding proteins** (**G-proteins**), which cleave guanosine triphosphate (GTP) to guanosine diphosphate (GDP), and depending on type (e.g. G_s, G_i, G_q), activate or inhibit other membrane-bound signalling enzymes such as **adenylyl cyclase**. Transmembrane proteins such as **integrins** and **cadherins** provide structural and signalling links with other cells and the **extracellular matrix** (Fig. 3b). Their cytosolic ends bind to components of the cytoskeleton, including protein kinases which can initiate, for example, altered gene transcription or changes in cell shape.

Mitochondria and energy production

Mitochondria use molecular oxygen to, in effect, burn sugar and small fatty acid molecules to produce **adenosine triphosphate (ATP)**, which is used by all energy-requiring cellular reactions. Glucose is first converted to pyruvate in the cytosol by glycolysis, producing in the process a small net amount of ATP and reduced nicotinic adenine dinucleotide (**NADH**). Glycolysis does not require O_2, so when O_2 is limited, this **anaerobic respiration** can supply some ATP, with NADH being reoxidized to NAD^+ by metabolism of the pyruvate to lactate (Fig. 3c). However, under normal conditions where there is sufficient O_2, **oxidative phosphorylation** in the mitochondria produces ~15 fold more ATP for each glucose molecule than does glycolysis. Pyruvate and fatty acids transported into the **mitochondrial matrix** act as substrates for enzymes that drive the **citric acid (Krebs') cycle**, which generates **NADH** and the waste product CO_2. The **electron transport chain**, a series of enzymes in the inner mitochondrial membrane, then uses molecular O_2 to re-oxidize NADH to NAD^+. In doing so, it generates a H^+ ion gradient across the inner membrane which drives the **ATP synthase** (Fig. 3c). Note that mitochondria are not solely devoted to ATP production, as they are also involved in other cellular processes, including Ca^{2+} homeostasis and signalling.

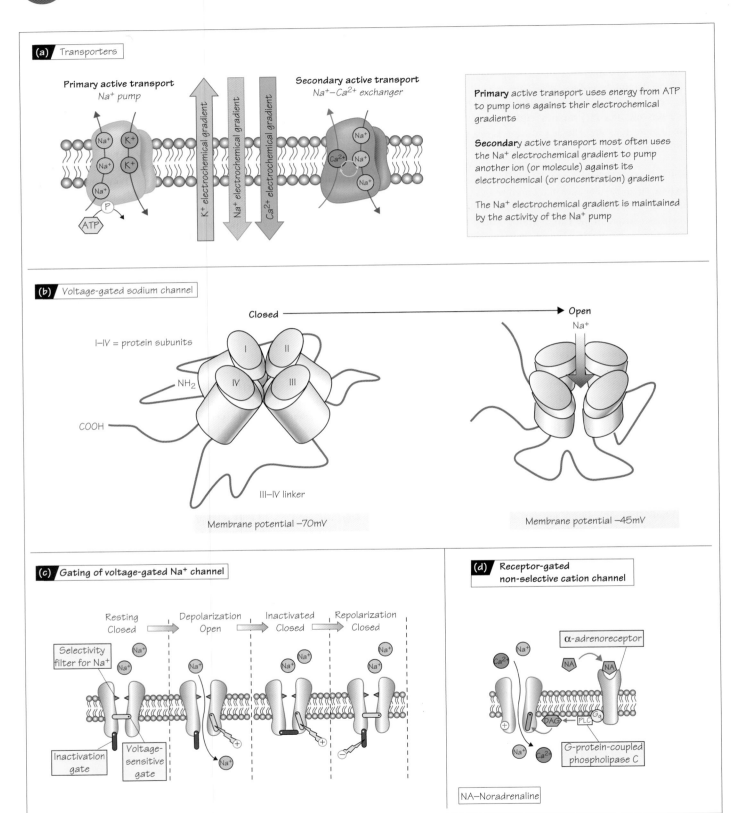

(a) Transporters

Primary active transport
Na⁺ pump

Secondary active transport
Na⁺–Ca²⁺ exchanger

K⁺ electrochemical gradient

Na⁺ electrochemical gradient

Ca²⁺ electrochemical gradient

Primary active transport uses energy from ATP to pump ions against their electrochemical gradients

Secondary active transport most often uses the Na⁺ electrochemical gradient to pump another ion (or molecule) against its electrochemical (or concentration) gradient

The Na⁺ electrochemical gradient is maintained by the activity of the Na⁺ pump

(b) Voltage-gated sodium channel

Closed ——————————→ Open
Na⁺

I–IV = protein subunits

NH₂

COOH

III–IV linker

Membrane potential –70mV

Membrane potential –45mV

(c) Gating of voltage-gated Na⁺ channel

Resting
Closed

Depolarization
Open

Inactivated
Closed

Repolarization
Closed

Selectivity
filter for Na⁺

Inactivation
gate

Voltage-
sensitive
gate

(d) Receptor-gated
non-selective cation channel

α-adrenoreceptor

G-protein-coupled
phospholipase C

NA–Noradrenaline

Physiology at a Glance, Third Edition. Jeremy P.T. Ward and Roger W.A. Linden.

© 2013 John Wiley & Sons, Ltd. Published 2013 by John Wiley & Sons, Ltd.

Proteins provide several routes for the movement of materials across membranes: (i) large pores, constructed of several protein subunits, that allow the bulk flow of water, ions and sometimes larger molecules (e.g. **aquaporin**, Chapter 34; and the **connexins**, that combine on the connexons to form **gap junctions** between cells); (ii) transporter molecules, some of which use metabolic energy (either direct or indirect) to move molecules against chemical and/or electrical gradients; and (iii) ion channels, specialized to allow the passage of particular ion species across the membrane under defined conditions.

Carrier-mediated transport

Transporter (or carrier) proteins can move a single type of molecule in one direction across a membrane (a **uniporter**), several different molecules in one direction (a **symporter**) or different molecules in opposite directions (an **antiporter**) (Fig. 4a). Transporters can allow the movement of molecules down chemical concentration gradients (**facilitated diffusion**), when the energy required for conformational changes in the transporter protein is provided by the concentration gradient rather than by metabolic activity. Important transporters for glucose and amino acids, found in the kidney and the gut, are in fact driven by the Na^+ electrochemical gradient that exists across the cell membrane (Chapter 2). These symporters must bind Na^+ *and* the primary transported molecule at the external surface of the membrane before the conformational change will take place. Antiporters such as the Na^+–Ca^{2+} exchanger similarly use the Na^+ gradient, in this case to extrude one Ca^{2+} out of the cell in exchange for three Na^+ into the cell. These processes are known as **secondary active transport**, as the Na^+ gradient is set up by a process requiring metabolic energy. The uneven distribution of Na^+ ions across the cell membrane is produced by the best known of all transporters, the **Na^+–K^+ ATPase**, also known as the **Na^+ pump** (Fig. 4a). This protein is an antiporter that uses metabolic energy to move Na^+ ions out of the cell and K^+ ions in, *against their respective concentration gradients*. The ATPase binds extracellular K^+ and intracellular Na^+ ions, usually in the ratio of 2:3, and hydrolyses adenosine triphosphate (ATP) to provide the energy needed to change its conformation, leading to the ejection of Na^+ into the extracellular medium and K^+ into the cytosol; this allows the cell to maintain a high concentration of K^+ ions and a low concentration of Na^+ ions inside the cell (Chapter 2). The Na^+ pump works continuously, although its activity is stimulated by high intracellular levels of Na^+ ions and can be modulated by second messenger-mediated phosphorylation. The action of the Na^+-K^+ ATPase is the most important example of **primary active transport**.

Ion channels

Ions can *diffuse* across cell membranes down their electrochemical gradient through **ion channels**. These transmembrane proteins, which are invariably constructed of several subunits containing several membrane-spanning domains (e.g. Fig. 4b), provide a charged, hydrophilic pore through which ions can move across the lipid bilayer. They possess a number of important features that confer upon the cell the ability to control closely the movement of ions across the membrane. Ion channels are **selective** for particular ions, i.e. they allow the passage of only one type of ion or a few related ions. There are numerous specialized channels for Na^+, K^+, Cl^- and Ca^{2+} ions, as well as non-specific channels for monovalent, divalent or even all cations (positively charged ions) or anions (negatively charged ions). The charge on the transmembrane pore determines whether the channel is for cations or anions, and selection between different ion types depends on the size of the ion and its accompanying water of hydration. Different types of channel for the same ion can however allow greatly differing amounts of that ion to move through them per second for the same electrochemical gradient; this is called channel **conductance**, and is best understood in the following way. Ions carry charge and so their movement causes an **electrical current**. Ohm's law states that V (voltage) = I (current) × R (resistance). In terms of ion channels, V = membrane potential and I = ionic current, so one can calculate the resistance of the channel. The reciprocal of resistance is conductance, which has units called Siemens; 1 Siemens (S) = 1/Ohm. Single ion channels generally have conductances in the 2–$300\,pS$ ($10^{-12}\,S$) range.

The second key feature of ion channels is that their pores are either **open** or **closed**; the transition between these states is called **gating**. Gating is brought about by a change in the conformation of the protein subunits that opens or closes the ion-permeable pore (e.g. Fig. 4b). Many channels are opened or closed according to the potential difference (voltage) across the cell membrane (**voltage** gating; Chapter 5), whereas others are gated by the presence of a specific signal molecule (**ligand** or **receptor** gating). The function of some channels may additionally be modified by phosphorylation of channel proteins by enzymes such as protein kinase C or A. The **voltage-gated fast inward Na^+ channel** that is responsible for the upstroke of the action potential (Chapter 5) has two gates, one that opens as the cell depolarizes beyond $\sim-55\,mV$ (its **threshold**) and another that shuts (inactivates) the channel as the potential becomes positive (Fig. 4c). This latter gate can only be reset by repolarizing towards the resting potential (Chapter 5). Some ligand-gated channels are directly gated by extracellular molecules, such as neurotransmitters or hormones, whereas others respond indirectly via intracellular signals, such as diacylglycerol (DAG; Fig. 4d) or cyclic adenosine monophosphate (cAMP) (Chapter 3). Specialized cells that detect changes in the internal and external environments (receptor cells) possess ion channels that are gated by the particular signal that is detected by the receptor, e.g. pH or light. The characteristics of ion channels, in concert with the activities of ion pumps, give cells the ability to control precisely the movement of ions across the cell membrane. This is crucial for many important physiological processes, including electrical signalling (Chapters 5 and 6), initiation of muscle contraction (Chapters 12 and 13) and the release of materials such as neurotransmitters, hormones and digestive enzymes.

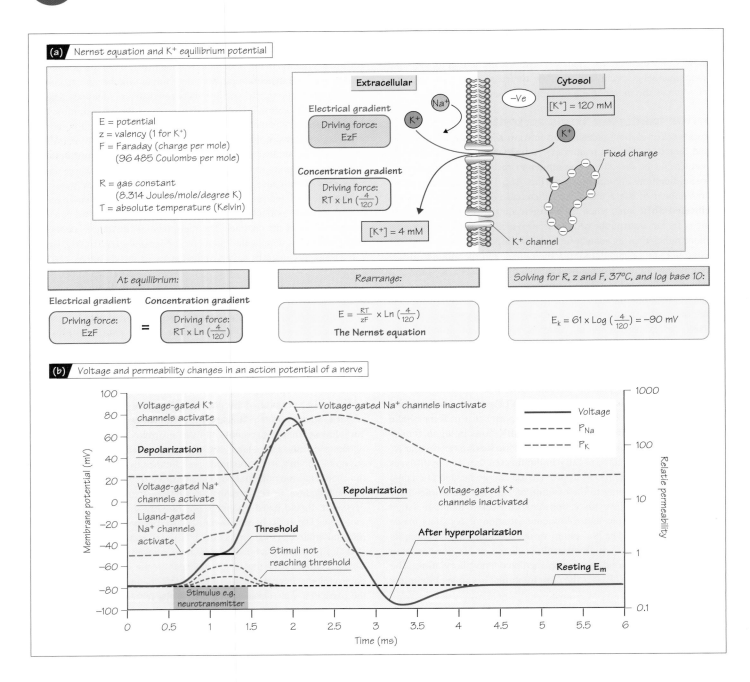

(a) Nernst equation and K⁺ equilibrium potential

E = potential
z = valency (1 for K⁺)
F = Faraday (charge per mole)
   (96 485 Coulombs per mole)

R = gas constant
   (8.314 Joules/mole/degree K)
T = absolute temperature (Kelvin)

Extracellular

Cytosol

$[K^+] = 120$ mM

Electrical gradient

Driving force:
EzF

Concentration gradient

Driving force:
$RT \times Ln \left(\frac{4}{120}\right)$

$[K^+] = 4$ mM

K⁺ channel

Fixed charge

At equilibrium:

Electrical gradient   Concentration gradient

Driving force:
EzF
=
Driving force:
$RT \times Ln \left(\frac{4}{120}\right)$

Rearrange:

$$E = \frac{RT}{zF} \times Ln \left(\frac{4}{120}\right)$$
The Nernst equation

Solving for R, z and F, 37°C, and log base 10:

$$E_k = 61 \times Log \left(\frac{4}{120}\right) = -90 \text{ mV}$$

(b) Voltage and permeability changes in an action potential of a nerve

Voltage-gated K⁺ channels activate

Voltage-gated Na⁺ channels inactivate

Depolarization

Voltage-gated Na⁺ channels activate

Ligand-gated Na⁺ channels activate

Threshold

Stimuli not reaching threshold

Repolarization

Voltage-gated K⁺ channels inactivated

After hyperpolarization

Resting E_m

Stimulus e.g. neurotransmitter

Legend: Voltage, P_{Na}, P_K

Membrane potential (mV): 100, 80, 60, 40, 20, 0, −20, −40, −60, −80, −100
Relative permeability: 1000, 100, 10, 1, 0.1
Time (ms): 0, 0.5, 1, 1.5, 2, 2.5, 3, 3.5, 4, 4.5, 5, 5.5, 6

Physiology at a Glance, Third Edition. Jeremy P.T. Ward and Roger W.A. Linden.

© 2013 John Wiley & Sons, Ltd. Published 2013 by John Wiley & Sons, Ltd.

Electrical events in biological tissues are caused by the movement of ions across the membrane. A potential difference exists across the membranes of all cells (**membrane potential**, E_m), but only **excitable tissues** can generate **action potentials** (transient depolarization of a cell as a result of ion channel activity). Action potentials transmit information in nerve cells (Chapter 6) and trigger contractions in muscle cells (Chapter 12). Cell membranes are electrically polarized so that the inside is negative relative to the outside. In excitable tissues, resting E_m is usually between –60 and –90 mV.

The resting membrane potential

The resting membrane is more permeable to K^+ and Cl^- than to other ions (Chapter 4). The cell contains negatively charged molecules (e.g. proteins) which cannot cross the membrane. This fixed negative charge attracts K^+, leading to accumulation of K^+ within the cell (Chapter 2). However, the consequent increase in the K^+ concentration gradient drives K^+ back out of the cell. This means fewer K^+ ions move into the cell than are required to achieve electrical neutrality with the fixed negative charges, and the inside of the cell therefore remains negatively charged compared to the outside, causing a potential difference across the membrane. Equilibrium is reached when the *electrical* forces exactly balance those due to *concentration* differences (**Gibbs–Donnan equilibrium**); the net force or **electrochemical gradient** for K^+ is then zero. If the membrane were *only* permeable to K^+, the voltage at which this would occur (**K^+ equilibrium potential**, E_K) is defined purely by the K^+ concentration gradient, and can be calculated from the **Nernst equation** (see Fig. 5a for derivation). Thus, if intracellular $[K^+]$ were 120 mmol/L and extracellular $[K^+]$ 4 mmol/L, $E_K = \sim$–90 mV. This applies to any ion, so if the membrane were *only* permeable to Na^+ (only Na^+ channels open) and intracellular and extracellular $[Na^+]$ were 10 and 140 mmol/L, respectively, the potential obtained at equilibrium (E_{Na}) would be +70 mV. To summarize, for any given intracellular and extracellular ionic concentrations, the equilibrium potential for that ion is the membrane potential required for the intracellular and extracellular concentrations to be in equilibrium, i.e. for the electrochemical gradient to be zero. The difference between the actual E_m and the equilibrium potential for any ion is therefore a measure of that ion's electrochemical gradient, the force driving it into or out of the cell.

Real cell membranes are permeable to other ions besides K^+, but at rest their K^+ permeability (P_K) is much greater than that for other ions. In particular, the ratio of P_K to Na^+ permeability (P_{Na}) ranges between 25 : 1 and 100 : 1 in nerve, skeletal and cardiac muscle cells. As a result E_m in such cells at rest (**resting membrane potential**) is close to E_K (–60 to –85 mV) and the electrochemical gradient for K^+ is small. E_m does not equal E_K because there is permeability to other ions, notably Na^+. As E_{Na} is much more positive than E_m, the Na^+ electrochemical gradient is strongly inwards, forcing Na^+ into the cell.

However, as P_{Na} is relatively low, only a small amount of Na^+ can leak in, though this is sufficient to slightly depolarize the membrane from E_K. A consequence of the above is that if P_{Na} were suddenly increased to more than P_K, then E_m would shift towards E_{Na}. This is exactly what happens during an action potential, when Na^+ channels open so that P_{Na} becomes 10-fold greater than P_K, and the membrane depolarizes.

The action potential

Action potentials are initiated in nerve and skeletal muscle by activation of **ligand-gated Na^+ channels** by neurotransmitters (Chapter 4 and 13). This increases P_{Na} and causes E_m to move towards E_{Na} (i.e. become positive; Fig. 5b). This initial increase in P_{Na} is however relatively modest, so the depolarization is similarly small. However, if the stimulus is sufficiently strong, E_m depolarizes enough to reach the **threshold potential** ($\sim$–55 mV), at which point **voltage-gated Na^+ channels** (Chapter 4) activate, causing further depolarization. This activates more voltage-gated Na^+ channels so the process becomes explosively self-regenerating, leading to a large transient increase in P_{Na} so it is 10-fold greater than P_K. As a result, E_m rapidly approaches E_{Na} ($\sim$+65 mV; see above), causing the sharp positive 'spike' or **depolarization** of the action potential, which lasts about 1 ms in nerve and skeletal muscle. The spike is transient because as E_m becomes positive, the voltage-gated Na^+ channels **inactivate** (Chapter 4) and P_{Na} plummets, whereas a type of voltage-gated K^+ channel (**delayed rectifier**) activates. Thus P_K is again much larger than P_{Na} and E_m returns towards E_K (**repolarization**); this takes about 1–2 ms. Delayed closure of the delayed rectifier K^+ channels means that the $P_K:P_{Na}$ ratio remains transiently greater than normal after repolarization, causing a transient hyperpolarization (Fig. 5b).

Following depolarization the Na^+ channels remain **inactive** for about 1 ms until the cell is largely repolarized and, during this period, they cannot be opened by any amount of depolarization. This is known as the **absolute refractory period** during which it is impossible to generate another action potential. For the following 2–3 ms, the transient hyperpolarization renders the cell more difficult to depolarize, an interval known as the **relative refractory period**, when an action potential can be generated only in response to a larger than normal stimulus. The refractory period limits the frequency at which action potentials can be generated to <1000/s and ensures that, once initiated, an action potential can travel only in one direction. Once triggered, an action potential will travel over the entire surface of an excitable cell (it is **propagated**) and will always have the same amplitude (it is **all-or-nothing**). The minute changes in ion concentrations that occur during an action potential are restored by the action of the Na^+ pump; it is important to understand that the action potential is *not* due to changes in ionic concentrations, but to changes in **ionic permeability**. Note that action potentials in cardiac muscle differ somewhat from those in nerves and skeletal muscle (Chapter 19).

6 Conduction of action potentials

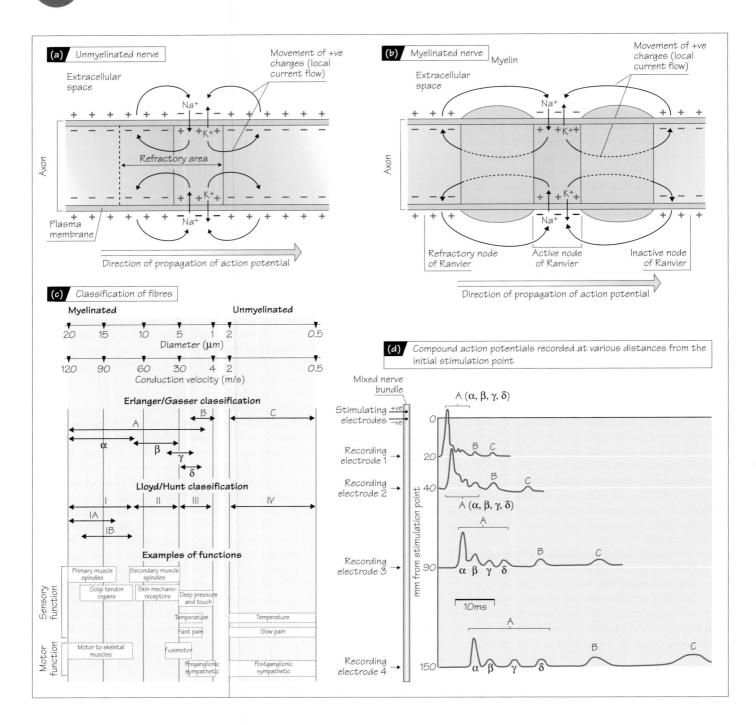

Physiology at a Glance, Third Edition. Jeremy P.T. Ward and Roger W.A. Linden.

© 2013 John Wiley & Sons, Ltd. Published 2013 by John Wiley & Sons, Ltd.

The **action potential** described in Chapter 5 is a local event that can occur in all excitable cells. This local event is an **all-or-nothing** response, leading to abolishion and then reversal of the polarity from negative (−70 mV) to positive (+40 mV) on the inside of the cell with respect to the outside for a short time during the course of the action potential.

Local currents are set up around the action potential because the positive charges from the membrane ahead of the action potential are drawn towards the area of negativity surrounding the action potential (current sink). This decreases the polarity of the membrane ahead of the action potential.

This electronic depolarization initiates a local response that causes the opening of the voltage-gated ion channels (Na^+ followed by K^+); when the threshold for firing of the action potential is reached, it **propagates** the action potential and this, in turn, leads to the local depolarization of the next area, and so on. Once initiated, an action potential does not depolarize the area behind it sufficiently to initiate another action potential because the area is **refractory** (Chapter 5).

This successive depolarization moves along each segment of an **unmyelinated** nerve until it reaches the end. It is all-or-nothing and does not decrease in size (Fig. 6a).

Saltatory conduction

Conduction in **myelinated** axons depends on a similar pattern of current flows. However, because **myelin** is an insulator and because the membrane below it cannot be depolarized, the only areas of the myelinated axon that can be depolarized are those that are devoid of any myelin, i.e. at the **nodes of Ranvier**. The depolarization jumps from one node to another and is called **saltatory**, from the Latin *saltare* (to jump) (Fig. 6b). Saltatory conduction is rapid and can be up to 50 times faster than in the fastest unmyelinated fibres.

Saltatory conduction not only increases the velocity of impulse transmission by causing the depolarization process to jump from one node to the next, but also **conserves energy** for the axon because depolarization only occurs at the nodes and not along the whole length of the nerve fibre, as in unmyelinated fibres. This leads to up to 100 times less movement of ions than would otherwise be necessary, therefore conserving the energy required to re-establish the Na^+ and K^+ concentration differences across the membranes following a series of action potentials being propagated along the fibre.

All nerve fibres are capable of conducting impulses in either direction if stimulated in the middle of their axon; however, normally they conduct impulses in one direction only (**orthodromically**), from either the receptor to the axon terminal or from the synaptic junction to the axon terminal. **Antidromic** conduction does not normally occur.

Fibre diameters and conduction velocities

Some information needs to be transmitted to and from the central nervous system very rapidly, whereas other information does not. Nerve fibres are able to cover both of these extremes and any in between by virtue of their size, and therefore conduction velocity, and whether or not they are myelinated. Nerve fibres come in all sizes, from 0.5 to 20 μm in diameter, with the smallest diameter unmyelinated fibres being the slowest conducting and the largest myelinated fibres the fastest conducting.

Classification of nerve fibres

Unfortunately, there are two classifications of nerve fibres. One, originally described by Erlanger and Gasser, and often referred to as the **general** classification, uses the letters **A, B and C, with A further subdivided into α, β, γ and δ**. The second, originally described by Lloyd and Hunt, and often referred to as the **sensory** or **afferent** classification, uses the Roman numerals **I, II, III and IV, with I further subdivided into A and B**. The groups are subdivided differently in the two classifications and so, unfortunately, it is not possible to rely on only one of the classifications for the description of nerve fibres. **The fibres of groups A and B and also of groups I, II and III are all myelinated, and those of group C and IV are unmyelinated.** These classifications, conduction velocities, fibre diameters and examples of their functions are shown in Figure 6c. A word of caution is necessary concerning the average conduction velocities of the larger myelinated fibres: in reality, although there may be a few larger diameter fibres in the human body that do indeed conduct impulses as fast as 120/ms, a more common observation is that the fastest proprioceptive (sensory) fibres conduct at below 100 m/s, with the average being closer to 80 m/s. The same applies to the α-motor neurones in that the conduction velocities rarely exceed 90 m/s, with the average being closer to 60 m/s.

Compound action potentials

Peripheral nerves in most animals comprise a number of axons bound together by a fibrous tissue called the **epineurium**. When extracellular recording electrodes are placed close to a peripheral nerve, the recorded voltage signal, when an action potential is initiated in the bundle, is much smaller (microvolts) than that recorded by an electrode inserted directly into the axon (millivolts). The extracellular recorded signal is made up of the electrical events occurring in all of the active fibres within the nerve bundle. If all the nerve fibres in a nerve bundle are synchronously stimulated at one end of a nerve, and recording electrodes are placed at a number of locations along the length of the bundle, a **compound action potential** is recorded at each electrode. The waveform recorded from each of the electrodes will differ due to the different conduction velocities of each group of fibres that makes up the bundle. Theoretically, if the nerve bundle were to contain examples of all classifications of nerve fibres (i.e. Aα, Aβ, Aγ, Aδ, B and C), the recorded compound action potential would be seen as a **multi-peaked** display, as the action potentials in the fastest conducting fibres (Aα) would reach the electrode before those in the slowest conducting fibres (C). Action potentials in the fibres with conduction velocities between these two extremes would arrive between these two times (Fig. 6d).

7 The autonomic nervous system

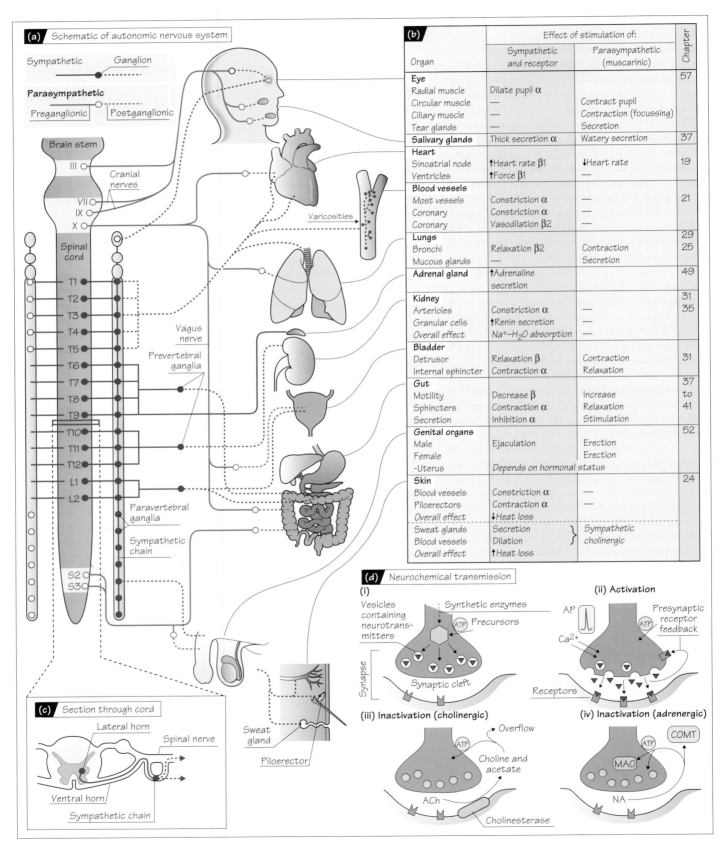

(a) Schematic of autonomic nervous system

Sympathetic — Ganglion

Parasympathetic

Preganglionic — Postganglionic

Brain stem

III — Cranial nerves

VII
IX
X

Spinal cord

T1
T2
T3
T4
T5
T6
T7
T8
T9
T10
T11
T12
L1
L2

S2
S3

Vagus nerve

Prevertebral ganglia

Paravertebral ganglia

Sympathetic chain

Varicosities

(b)

Organ	Effect of stimulation of:		Chapter
	Sympathetic and receptor	Parasympathetic (muscarinic)	
Eye			57
Radial muscle	Dilate pupil α		
Circular muscle	—	Contract pupil	
Ciliary muscle	—	Contraction (focussing)	
Tear glands		Secretion	
Salivary glands	Thick secretion α	Watery secretion	37
Heart			
Sinoatrial node	↑Heart rate β1	↓Heart rate	19
Ventricles	↑Force β1	—	
Blood vessels			
Most vessels	Constriction α	—	21
Coronary	Constriction α	—	
Coronary	Vasodilation β2	—	
Lungs			29
Bronchi	Relaxation β2	Contraction	25
Mucous glands	—	Secretion	
Adrenal gland	↑Adrenaline secretion		49
Kidney			31
Arterioles	Constriction α	—	35
Granular cells	↑Renin secretion	—	
Overall effect	Na⁺–H₂O absorption	—	
Bladder			
Detrusor	Relaxation β	Contraction	31
Internal sphincter	Contraction α	Relaxation	
Gut			37
Motility	Decrease β	Increase	to
Sphincters	Contraction α	Relaxation	41
Secretion	Inhibition α	Stimulation	
Genital organs			52
Male	Ejaculation	Erection	
Female		Erection	
-Uterus	Depends on hormonal status		
Skin			24
Blood vessels	Constriction α	—	
Piloerectors	Contraction α	—	
Overall effect	↓Heat loss	—	
Sweat glands	Secretion	} Sympathetic cholinergic	
Blood vessels	Dilation		
Overall effect	↑Heat loss		

(d) Neurochemical transmission

(i)

Vesicles containing neurotransmitters — Synthetic enzymes

ATP Precursors

Synapse

Synaptic cleft

(ii) Activation

AP

Presynaptic receptor feedback

Ca²⁺

ATP

Receptors

(iii) Inactivation (cholinergic)

Overflow

ATP

Choline and acetate

ACh

Cholinesterase

(iv) Inactivation (adrenergic)

COMT

ATP

MAO

NA

(c) Section through cord

Lateral horn

Spinal nerve

Ventral horn

Sympathetic chain

Sweat gland

Piloerector

Physiology at a Glance, Third Edition. Jeremy P.T. Ward and Roger W.A. Linden.

© 2013 John Wiley & Sons, Ltd. Published 2013 by John Wiley & Sons, Ltd.

The **autonomic nervous system** (**ANS**) provides the **efferent** pathway for the **involuntary** control of most organs, excluding the motor control of skeletal muscle (Chapters 12 and 13). The ANS provides the effector arm for **homeostatic reflexes** (e.g. control of blood pressure), and allows the integration and modulation of function by central mechanisms in the brain in response to *environmental* and *emotional* stimuli (e.g. exercise, thermoregulation, 'fight or flight'). Figure 7a shows a simplified schematic diagram of the ANS, and Figure 8b its actions on major organs.

The ANS is divided into **sympathetic** and **parasympathetic** systems. Both contain **preganglionic neurones** originating in the central nervous system that synapse with non-myelinated **postganglionic neurones** in the **peripheral ganglia**; postganglionic neurones innervate the target organ or tissue (Fig. 7a,b). Preganglionic neurones of both sympathetic and parasympathetic systems release **acetylcholine** in the synapse, which acts on **cholinergic nicotinic** receptors on the postganglionic fibre. The postganglionic neurotransmitters and receptors depend on the system and organ (see below). Parasympathetic peripheral ganglia are generally found close to or in the target organ, whereas sympathetic ganglia are largely located in two **sympathetic chains** either side of the vertebral column (*paravertebral ganglia*), or in diffuse *prevertebral ganglia* of the visceral plexuses of the abdomen and pelvis (Fig. 7a). Sympathetic postganglionic neurones are therefore generally long, whereas parasympathetic neurones are generally short. An exception is the sympathetic innervation of the **adrenal gland**, where preganglionic neurones directly innervate the adrenal medulla.

The sympathetic system is more pervasive than the parasympathetic; where an organ is innervated by both systems, they often act antagonistically (Fig. 7b). However, there is a high degree of central coordination, so that an increase in sympathetic activity to an organ is commonly accompanied by a decrease in parasympathetic activity. Sympathetic and parasympathetic activity may modulate different functions in the same organ (e.g. genital organs). In loose terms, the sympathetic system might be said to coordinate '*flight or fight*' responses, and the parasympathetic system '*rest and digest*' responses.

Sympathetic system

Sympathetic preganglionic neurones originate in the *lateral horn* of segments T1–L2 of the spinal cord, and exit the cord via the *ventral horn* (Fig. 7c) on their way to the paravertebral or prevertebral ganglia. Sympathetic postganglionic neurones terminate in the effector organs, where they release **noradrenaline** (**norepinephrine**). Noradrenaline and **adrenaline** (**epinephrine**), which is released by the adrenal medulla, are catecholamines, and activate **adrenergic** receptors, which are linked via G-proteins to cellular effector mechanisms. There are two main classes of adrenergic receptor, α and β, and these are further subdivided into several subtypes (e.g. α_1, α_2, β_2, β_2). Noradrenaline and adrenaline are equally potent on α_1-receptors, which are linked to G_q-proteins and are commonly associated with smooth muscle contraction (e.g. blood vessels). The α_2-receptors are $G_{i/o}$-protein linked and are often inhibitory. All β-receptors are linked to G_s-protein and activate adenylyl cyclase to make cyclic adenosine monophosphate (cAMP). Noradrenaline is more potent at β_1-receptors and adrenaline is more potent at β_2-receptors. The activation of β-receptors is associated with the relaxation of smooth muscle (e.g. blood vessels, airways), but it increases heart rate and force (Fig. 7b).

A few sympathetic neurones release acetylcholine at the effector (e.g. sweat glands), and are thus known as **sympathetic cholinergic** neurones.

Parasympathetic system

Parasympathetic preganglionic neurones originate in the brain stem, from which they run in cranial nerves III, VII, IX and X (**vagus**), and also from the second and third sacral segments of the spinal cord (Fig. 7a). Parasympathetic postganglionic neurones release acetylcholine, which acts on **cholinergic muscarinic** receptors. Parasympathetic activation causes secretion in many glands (e.g. bronchial mucous glands), and either contraction (e.g. bladder detrusor) or relaxation (e.g. bladder internal sphincter) of smooth muscle, although it has little effect on blood vessels. Notable exceptions, however, include vasodilatation in the penis and clitoris with subsequent erection (Chapter 51).

Neurochemical transmission

Action potentials (APs) in incoming neurones are transmitted by the release of neurotransmitters that bind to receptors on the postganglionic neurone or effector tissue. Between neurones (e.g. in ganglia), this occurs within a classical **synapse**, where the axon terminates in a bulbous swelling or **bouton** separated from the target by a narrow (10–20 nm) synaptic cleft (Fig. 7di). Postganglionic neurones branch repeatedly and have numerous boutons along their length, forming **varicosities** (e.g. see blood vessel in Fig. 7a). The boutons may either be close (~20 nm) to the effector membrane, allowing fast and specific delivery of the signal, or at some distance (100–200 nm), allowing a more distributed but slower effect. The mechanisms of neurochemical transmission are similar, and although the text below and Fig. 7di–iv refer to synapses, the same principles apply.

Synthetic enzymes are transported down the axon into the bouton, where they synthesize neurotransmitter (acetylcholine, noradrenaline) from precursors transported into the bouton. The neurotransmitter is stored in 50-nm **vesicles** (Fig. 7di). The arrival of an AP at the nerve ending causes an influx of Ca^{2+}, the fusion of vesicles with the membrane and the release of neurotransmitter; this binds to postsynaptic receptors and activates the response. Neurotransmitter release can be suppressed by feedback onto **presynaptic inhibitory receptors** (α_2-*receptors* for adrenergic synapses) (Fig. 7dii). Neurotransmitters must be removed at the end of activation. In *cholinergic* synapses, **cholinesterase** rapidly breaks down acetylcholine into *choline* and *acetate*, which are recycled; some may escape into interstitial fluid (*overflow*) (Fig. 7diii). In adrenergic synapses, most noradrenaline is rapidly taken up again by the nerve ending via an adenosine triphosphate (ATP)-dependent transporter called **uptake-1**; recovered noradrenaline is recycled. Some facilitated diffusion (**uptake-2**) also occurs into smooth muscle. Excess noradrenaline and sympathomimetic amines, such as tyramine (found in some foodstuffs), are metabolized in the neurone by mitochondrial **monoamine oxidase** (**MAO**). Noradrenaline and other catecholamines that enter the circulation are metabolized sequentially by **catechol-O-methyl transferase** (**COMT**) and MAO (Fig. 7div).

8 Blood

(a) Blood cell counts and haematocrit

Erythrocytes	male	4.7–6.1	$\times 10^{-12}$/L
	female	4.2–5.4	
Packed cell volume (PCV)	male	0.41–0.52	(no unit)
(Haematocrit)	female	0.36–0.48	
Haemoglobin (Hb)	male	130–180	g/L
	female	120–160	
Leucocytes (total)		4–11	$\times 10^9$/L
(white blood cell count, WBCC)			
Platelets		150–400	$\times 10^9$/L

(b) Plasma proteins

	Content (g/L)	Typical functions
Albumin	48	Oncotic pressure Binds hormones and drugs
α-globulins	5.5	Copper transport Antiproteases
β-globulins:		
Transferrin	3	Iron transport
Prothrombin	1	Haemostasis
Plasminogen	0.7	Haemostasis
Complement	1.6	Immune system
Fibrinogen	3	Haemostasis
γ-globulins	13	Immune system

(c) Erythropoiesis and the life cycle of the erythrocyte

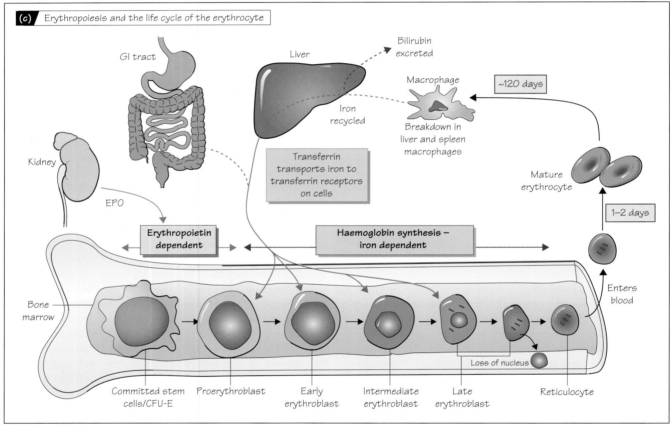

(d) White blood cells (numbers and proportions change greatly in disease)

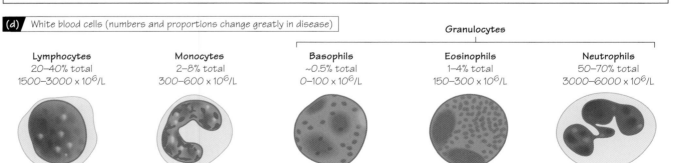

Granulocytes

Lymphocytes
20–40% total
1500–3000 $\times 10^6$/L

Monocytes
2–8% total
300–600 $\times 10^6$/L

Basophils
~0.5% total
0–100 $\times 10^6$/L

Eosinophils
1–4% total
150–300 $\times 10^6$/L

Neutrophils
50–70% total
3000–6000 $\times 10^6$/L

Physiology at a Glance, Third Edition. Jeremy P.T. Ward and Roger W.A. Linden.

© 2013 John Wiley & Sons, Ltd. Published 2013 by John Wiley & Sons, Ltd.

The primary function of blood is to deliver O_2 and energy to the tissues, and remove CO_2 and waste products. It is also important for the defence and immune systems, regulation of temperature, and transport of hormones and signalling molecules between tissues. Blood consists of **plasma** (Chapter 2) and blood cells. **Red blood cells** contain haemoglobin and transport respiratory gases (Chapter 28), whereas **white cells** form part of the defence system (Chapter 10). In adults, all blood cells are produced in the red bone marrow. Normal values for cell counts, haemoglobin and proportion of blood volume due to red cells (**haematocrit** or **packed cell volume**; estimated by centrifuging a blood sample) are shown in Figure 8a. **Platelets** are discussed in Chapter 9.

Plasma proteins

Plasma contains several important proteins (Fig. 8b), with a total concentration of 65–83 g/L. Most, other than γ-globulins (see below), are synthesized in the liver. Proteins can ionize as either acids or bases because of the presence of both NH_2 and COOH groups. At pH 7.4 they are mostly in the anionic (acidic) form. Their ability to accept or donate H^+ means they can act as **buffers** (Chapter 36). Plasma proteins have important **transport functions**, as they bind many hormones (e.g. cortisol and thyroxine) and metals (e.g. iron). They are classified into **albumin**, **globulin** and **fibrinogen** fractions. Globulins are further classified as **α-, β- and γ-globulins**. Examples and their major functions are shown in Figure 8b.

Red blood cells

Red blood cells (**erythrocytes**) are biconcave discs ~8 μm wide, and uniquely have no nucleus. They therefore cannot repair themselves and have a lifespan of only 100–120 days. The shape and flexibility of red cells allows them to deform easily and pass through capillaries. Importantly, they contain **haemoglobin** which is responsible for carriage of O_2, and also plays a role in acid–base buffering (Chapter 28 and 36).

Red cells are formed by a process called **erythropoiesis** (Fig. 8c). They originate from **committed stem cells** in the bone marrow of the adult, and liver and spleen of the fetus. The glycoprotein hormone **erythropoietin** (EPO) increases the number of committed stem cells and promotes production of red cells. Erythropoietin is produced mainly by the kidneys in adults, and liver in the fetus. The key stimulus for increased erythropoietin is low O_2 (**hypoxia**). Stem cells differentiate into **erythroblasts** (**early normoblasts**), which are relatively large (~15 μm) and nucleated. As differentiation proceeds, the cells shrink and **haemoglobin** is synthesized, which requires **iron**, **folate** and **vitamin B_{12}**. In the **late normoblast** the nucleus breaks up and disappears. The young red cell shows a reticulum on staining, and is called a **reticulocyte**. As it ages, the reticulum disappears and the characteristic biconcave shape develops. Normally 1–2% of circulating red cells are reticulocytes. This increases when erythropoiesis is enhanced (e.g. by hypoxia). About 2×10^{11} red cells are produced from the marrow each day. The spleen holds a reserve of red cells that can be released following blood loss.

Red cells are destroyed by **macrophages** in the liver and spleen after ~120 days. The haem group is split from haemoglobin and converted to **biliverdin** and then **bilirubin**. The iron is conserved and recycled via **transferrin**, an iron transport protein, or stored in **ferritin**. Bilirubin is a brown–yellow compound which is excreted in the bile. An increased rate of haemoglobin breakdown results in excess bilirubin, which stains the tissues (**jaundice**).

An inadequate amount of red cells and/or haemoglobin is called **anaemia**. This is commonly a result of **haemorrhage** (e.g. heavy menstruation), but also occurs when the diet contains insufficient iron, folate or vitamin B_{12}, or they are poorly absorbed in the gut (Chapter 39). Anaemia is also caused by abnormalities of haemoglobin (**thalassaemia**), the **sickle cell** mutation and **leukaemia** (white cell cancers).

Red cells have surface **antigens** that can react with specific antibodies in the plasma. The antigens and antibodies present are determined genetically, forming the basis of **blood groups**. The most important systems are **ABO** (A, B, both or neither antigens present) and **Rh** (Rhesus; D or no D antigen). Matching of blood groups is essential during blood transfusions, because red cells with a different antigen to the recipient will react with antibodies in the plasma, stick together (**agglutinate**) and **haemolyse** (break apart). The Rh system is important in pregnancy, because an Rh– mother can be sensitized (produce antibodies) to red cells from a Rh+ fetus during birth. This can be a problem for a second pregnancy with another Rh+ fetus, as antibodies cross the placenta.

White blood cells

White blood cells (**leucocytes**) defend the body against infection by foreign material, and the white cell count (Fig. 8a) increases greatly in disease. Three main types are present in blood: granulocytes, lymphocytes and monocytes.

Granulocytes are further classified as **neutrophils** (neutral-staining granules), **eosinophils** (acid-staining granules) and **basophils** (basic-staining granules) (Fig. 8d). All contribute to inflammation by releasing mediators. **Neutrophils** have a key role in the **innate immune system**, and migrate to areas of infection (**chemotaxis**) within minutes, where they destroy bacteria by **phagocytosis** (engulfing them). They are a major component of pus. Neutrophils live for ~6 h in blood, longer in tissues. **Eosinophils** are less motile but longer lived, and phagocytose larger parasites. They are increased in allergic disease, to which they contribute by releasing inflammatory mediators. **Basophils** release histamine and heparin as part of the inflammatory response and are similar to tissue **mast cells**.

Lymphocytes originate in the bone marrow but mature in the lymph nodes, thymus and spleen before returning to the circulation. Most remain in the lymphatic system. Lymphocytes are critical components of the **immune system** and are of three main forms: **B cells** which produce γ-globulins (**immunoglobulins,** antibodies), **T cells** which coordinate the immune response, and **natural killer** (NK) cells which kill infected or cancerous cells (Chapter 10).

Monocytes are phagocytes but larger and longer lived than granulocytes. After formation in the marrow they circulate in the blood for ~72 h before entering tissues to become **macrophages**, which unlike granulocytes can also dispose of dead cell debris. Macrophages form the **reticuloendothelial system** in liver, spleen and lymph nodes.

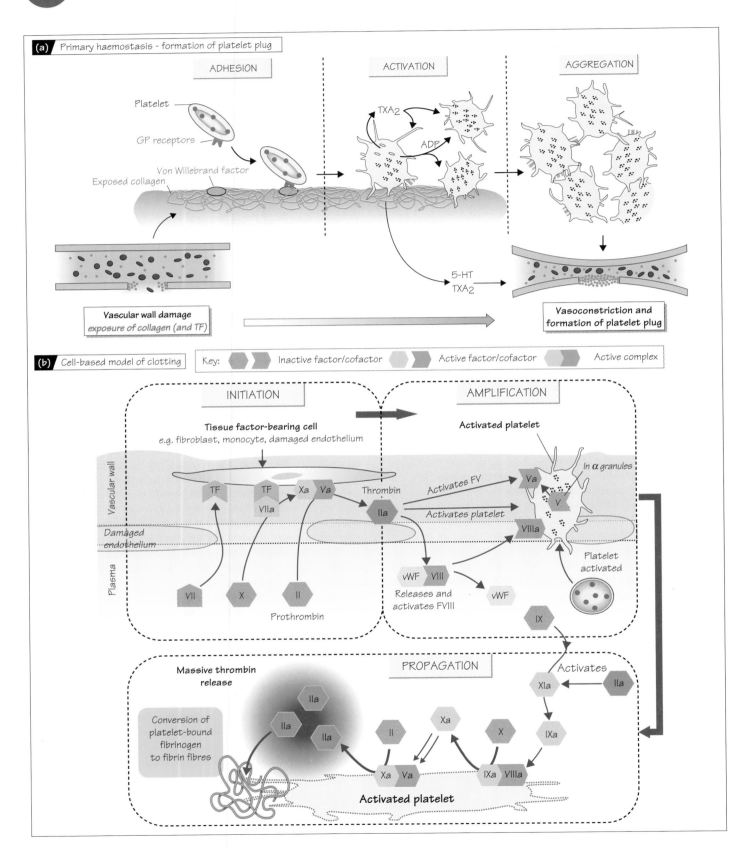

Physiology at a Glance, Third Edition. Jeremy P.T. Ward and Roger W.A. Linden.

 © 2013 John Wiley & Sons, Ltd. Published 2013 by John Wiley & Sons, Ltd.

Leaks in the cardiovascular system can lead to loss of blood and must be rapidly plugged. This is the purpose of **haemostasis**, a complex process that includes formation of the blood clot, a tough mesh of **fibrin** entrapping platelets and blood cells. Its complexity is in part due to the precarious balance that must be maintained between providing a rapid and effective means of stopping leaks, and inappropriate formation of clots in blood vessels (**thrombosis**). Thrombosis is associated with many serious conditions such as coronary artery disease.

Platelets play a critical role in haemostasis. They circulate in the blood but are not true cells, being small (~3 μm) vesicle-like structures formed from megakaryocytes in the bone marrow. They have a lifespan of ~4 days. Platelets have multiple surface receptors and clearly visible dense granules. These contain mediators such as serotonin and adenosine diphosphate (ADP), which are released on activation.

Primary haemostasis

The immediate response to damage of a blood vessel wall is **vasoconstriction**, which reduces blood flow and thus loss; it is an intrinsic property of the blood vessel. This is followed by a sequence of events that eventually leads to sealing of the wound by a clot (Fig. 9a). Damage to the vessel wall exposes collagen, to which a plasma protein called **von Willebrand factor** (vWF) binds. Tissue factor (TF) is also exposed (see below). **Platelets** have **glycoprotein** (GP) receptors which avidly bind to vWF, tethering the platelet. Further receptors including **integrins** (Chapter 3) bind directly to collagen. Together these cause **adhesion** of the platelet to the damaged area. Binding to these receptors also initiates **platelet activation**, partly by increasing intracellular Ca^{2+}. Platelets change shape, put out pseudopodia and make **thromboxane A_2** (TXA_2) via **cyclooxygenase** (COX). TXA_2 stimulates release of **serotonin**, **ADP** and other compounds from the platelet granules. TXA_2 and serotonin also enhance the vasoconstriction. The process propagates because **ADP** directly activates more platelets via puringeric (**P2Y**) receptors. It also causes activation of fibrinogen (**GPIIb/IIIa**) receptors on their surface, which bind to **fibrinogen** in the plasma causing the platelets to become sticky and **aggregate**, forming a soft platelet plug (Fig. 9a). This is stabilized during clotting by conversion of the fibrinogen to **fibrin**. Note that **thrombin** (see below) is also a potent platelet activator.

Formation of the blood clot (coagulation)

The process leading to formation of the blood clot involves sequential conversion of **proenzymes** to active enzymes (**clotting factors**; e.g. factor X → Xa) in an amplifying cascade. Most clotting factors are produced in the liver, which requires **vitamin K**, and many (e.g. thrombin, factor X) require Ca^{2+} to act. The ultimate purpose is to produce a massive burst of **thrombin** (factor IIa; a protease which cleaves fibrinogen to fibrin), and thus rapid formation of the clot. The **cell-based model of clotting** (Fig. 9b) has replaced the older extrinsic and intrinsic pathways. Most of the action in this model only occurs on cell or platelet surfaces (hence its name).

The **initial phase** of clotting occurs when cells that express a protein called **tissue factor** (**TF**; thromboplastin) become exposed to plasma as a result of vascular damage. Such cells include fibroblasts, monocytes and damaged endothelial cells. **Factor VIIa** from the plasma is then able to bind to TF (**TF : VIIa**), and this complex consequently activates a key protein in the clotting process, **factor X**. It is this, when combined with **cofactor Va** to form **prothrombinase**, that converts **prothrombin** to **thrombin**; importantly, it can only do so when tethered to the cell by phospholipids. However, insufficient thrombin is produced at this stage for clot formation, and the signal has to be amplified.

The **amplification phase** takes place on the surface of **platelets** (Fig. 9b). The small amount of thrombin produced above activates nearby platelets, and also cofactor V on their surface. **Cofactor VIII** is normally bound to plasma vWF, which protects it from degradation. Thrombin cleaves factor VIII from vWF and activates it, when it also binds to the platelet surface. The end product is a large number of activated platelets, each with a large surface area (due to pseudopodia) covered with the active cofactors, all stuck together by fibrinogen (see above).

The scene is now set for the **propagation phase**. Thrombin activates a short cascade that leads to activation of **factor IX** (also activated by TF:VIIa). Factor IXa forms a complex with cofactor VIIIa on the platelet surface to form **tenase**, a much more powerful activator of factor X than TF:VIIa. The large amount of factor Xa thus generated binds to cofactor Va also on the platelet surface to form a similarly large amount of **prothrombinase**. There is consequently a massive burst of thrombin production, 1000-fold greater than in the initial phase. Thrombin cleaves the **fibrinogen** bound around the platelets to form **fibrin monomers**, which spontaneously polymerize to a fibrous mesh of fibrin, entrapping the platelets and other blood cells. The fibrin polymer is finally cross-linked by **factor XIIIa** (also activated by thrombin) to create a tough network of fibrin fibres and a stable clot. Retraction of entrapped platelets contracts the clot by ~60%, making it tougher and assisting repair by drawing the edges of the wound together.

Inhibitors of haemostasis and fibrinolysis

Because thrombin both activates and is produced by the mechanisms described above, there is an element of **positive feedback**, and the whole process is intrinsically unstable. Multiple inhibitory mechanisms counteract this to prevent inappropriate clotting. Undamaged endothelium produces **prostacyclin** and **nitric oxide** (Chapter 24), which impede platelet adhesion and activation and so limit them to damaged areas. Plasma **antithrombin** inhibits thrombin, factor Xa and tenase, and is strongly potentiated by **heparin** and **heparans** on endothelial cells. **Thrombomodulin** (also on endothelial cells) binds thrombin and converts it so it no longer cleaves fibrinogen but instead activates **protein C** (APC), which with **protein S** inactivates factors Va and VIIIa, and hence tenase and prothrombinase. Finally, the clot is broken down by **plasmin**, a process called **fibrinolysis**. This occurs when plasma **plasminogen** binds to fibrin, and is converted to plasmin by **tissue plasminogen activator** (tPA). Plasmin is itself inactivated by α_2-antiplasmin.

10 Defence: Inflammation and immunity

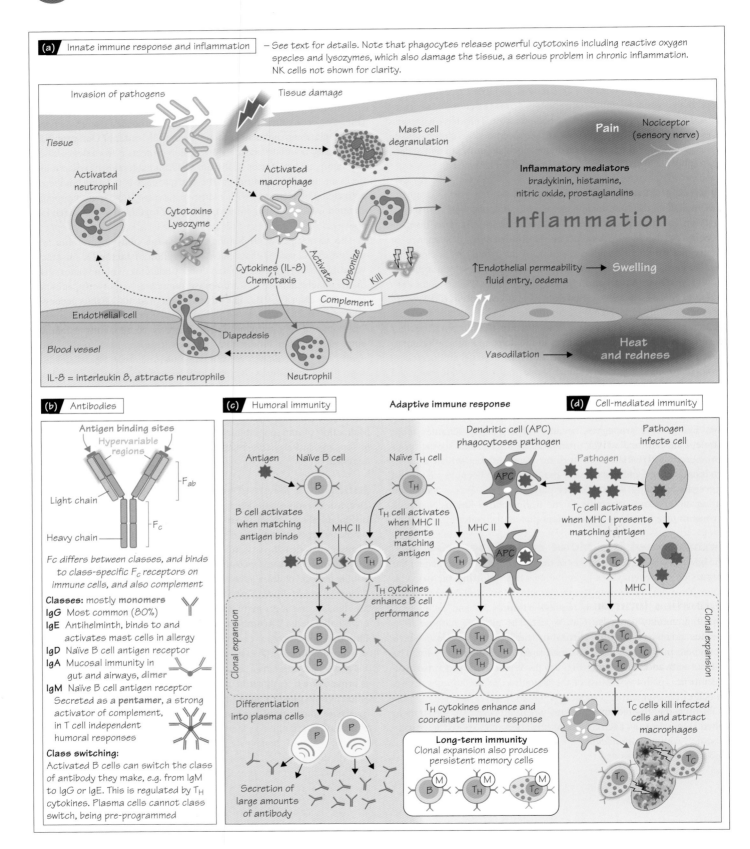

(a) Innate immune response and inflammation — See text for details. Note that phagocytes release powerful cytotoxins including reactive oxygen species and lysozymes, which also damage the tissue, a serious problem in chronic inflammation. NK cells not shown for clarity.

Invasion of pathogens

Tissue damage

Tissue

Activated neutrophil

Cytotoxins Lysozyme

Activated macrophage

Mast cell degranulation

Pain   Nociceptor (sensory nerve)

Inflammatory mediators
bradykinin, histamine, nitric oxide, prostaglandins

Inflammation

Cytokines (IL-8) Chemotaxis

Activate   Opsonize   Kill

Complement

↑Endothelial permeability fluid entry, oedema → Swelling

Endothelial cell

Diapedesis

Blood vessel

Neutrophil

Vasodilation → Heat and redness

IL-8 = interleukin 8, attracts neutrophils

(b) Antibodies

Antigen binding sites
Hypervariable regions

F_{ab}

Light chain

F_c

Heavy chain

Fc differs between classes, and binds to class-specific F_c receptors on immune cells, and also complement

Classes: mostly **monomers**
IgG Most common (80%)
IgE Antihelminth, binds to and activates mast cells in allergy
IgD Naïve B cell antigen receptor
IgA Mucosal immunity in gut and airways, **dimer**
IgM Naïve B cell antigen receptor Secreted as a **pentamer**, a strong activator of complement, in T cell independent humoral responses

Class switching:
Activated B cells can switch the class of antibody they make, e.g. from IgM to IgG or IgE. This is regulated by T_H cytokines. Plasma cells cannot class switch, being pre-programmed

Adaptive immune response

(c) Humoral immunity

Antigen   Naïve B cell   Naïve T_H cell

B cell activates when matching antigen binds

T_H cell activates when MHC II presents matching antigen

MHC II

T_H cytokines enhance B cell performance

Clonal expansion

Differentiation into plasma cells

Secretion of large amounts of antibody

Dendritic cell (APC) phagocytoses pathogen

APC

MHC II

Long-term immunity
Clonal expansion also produces persistent memory cells

T_H cytokines enhance and coordinate immune response

(d) Cell-mediated immunity

Pathogen infects cell

Pathogen

T_C cell activates when MHC I presents matching antigen

MHC I

Clonal expansion

T_C cells kill infected cells and attract macrophages

Physiology at a Glance, Third Edition. Jeremy P.T. Ward and Roger W.A. Linden.

30 © 2013 John Wiley & Sons, Ltd. Published 2013 by John Wiley & Sons, Ltd.

Physical defence against infection by **bacteria, viruses, fungi,** and **parasites** is provided by the skin, and epithelia lining the airways and gut. The latter secrete **anti-microbial** chemicals and **mucus**, which traps microorganisms and is removed by cilia (Chapter 25) or peristalsis. Haemostasis quickly seals breaches (Chapter 9). Organisms evading these defences are targeted by the **immune system**, where **leucocytes** play a central role (Chapter 8). The **innate** immune response is fast but *non-specific* and causes **inflammation**, characterised by **heat, redness, swelling** and **pain**. The **adaptive** response is slower, *highly specific*, and more potent.

Innate immune response

Tissue damage and invasion of pathogens activate **mast cells** (similar to *basophils*) and resident **phagocytes**, primarily **macrophages** and **dendritic cells**, which release **inflammatory mediators**, signalling molecules (**cytokines**) and **cytotoxic agents** (Fig. 10a). Inflammatory mediators cause **vasodilation** (*heat and redness*), stimulate **nociceptors** (*pain*) and increase **endothelial permeability**, leading to *extravasation* of protein and fluids and thus *oedema* (*swelling*) (Chapters 2 and 23). Cytokines (e.g. interleukin 8, IL-8) attract many more phagocytes, chiefly **neutrophils** (**chemotaxis**); these leave the blood by squeezing between endothelial cells (**diapedesis**). Phagocytes ingest (**phagocytose**) microorganisms, and in the case of macrophages also damaged cells and debris. Pathogens can be detected because they express **pathogen-associated molecular patterns** (**PAMPs**) not found in mammals (e.g. bacterial mannose residues, viral RNA and fungal glucans). PAMPs are recognised by phagocyte **pattern recognition receptors** (**PRRs**); relatively few different PRRs are required (<1000) because PAMPs are common across wide groups of pathogen. On binding a PAMP, PRRs initiate phagocytosis and release of cytokines and cytotoxins. Injured, infected or cancerous cells express PAMP-like molecules recognised by **natural killer** (**NK**) lymphocytes, which kill the cells and activate macrophages to remove the debris. In major infections cytokines such as IL-1 cause *fever*; high temperatures may assist the immune response.

Complement is an important *non-cellular* mechanism comprised of a cascade of plasma proteins. On activation it coats and **opsonizes** (*facilitates phagocytosis*) pathogens, kills by membrane rupture, recruits phagocytes and induces inflammation. It is activated by some surface molecules (e.g. bacterial mannose) and by *antibodies* (e.g. IgM; Fig. 10b) that have 'tagged' a pathogen or material as foreign.

Antibodies (immunoglobulins)

Adaptive immunity depends on **antibodies**, which are made by **lymphocytes** and recognise highly specific molecular sequences (**epitopes**) on proteins, polysaccharides, lipids and small chemicals. Molecules that react with antibodies are called **antigens**. There are five antibody classes (Fig. 10b). All have a constant region (F_C) attached to two *hyper-variable* branches (F_{ab}) which recognise the epitope. The hypervariability is due to random mutations in antibody genes during lymphocyte maturation, so each cell can end up with one of $\sim10^9$ different antibodies. Although *individual cells express just one variant*, the large number of lymphocytes and random nature of production means that every variant will be expressed somewhere, if only in a small group of cells. Such groups of lymphocytes with identical antibodies are called **clones**. Any lymphocytes with antibodies directed against *self* are (normally) destroyed during maturation. Antibodies *neutralise* toxins and prevent attachment of pathogens; *target, opsonize* or *agglutinate* (clump together) antigens for phagocytosis; *target* pathogens and foreign material for complement; and, crucially, act as **antigen receptors** on lymphocytes.

Adaptive immune response

The adaptive response takes ~5 days to become effective, and peaks after 1–2 weeks. It has two intertwined branches: **humoral immunity**, mediated by **B lymphocytes** (B cells) which mature in **bone marrow**, and **cell-mediated immunity**, mediated by **T lymphocytes** (T cells) which mature in the **thymus**. **Naïve** (not yet activated) lymphocytes continually **recirculate** between **lymphoid tissues** (e.g. lymph nodes, tonsils and spleen) until they encounter a matching antigen.

Humoral immunity (Fig. 10c) is particularly effective against extracellular pathogens, as it involves secretion of antibodies into extracellular fluid. Only B cells can do this, or have antigen receptors that can recognise *all* types of antigen (e.g. protein, polysaccharide, lipid, etc.). When an antigen binds to its matching receptor on naïve B cells, the latter activate and undergo **clonal expansion** – rapid proliferation resulting in a large number of identical cells expressing the same antibody. These differentiate into **plasma cells**, which secrete the antibody in massive amounts. For non-protein antigens the whole process is **T cell independent**. However, if the antigen is a protein, **T helper** (**T_H**, CD4+) cells substantially enhance the response. T cells only recognise protein (or peptide) antigens, and then only when they are *presented* to them by **major histocompatibility complex** (**MHC II**) on **antigen presenting cells** (**APC**), which include dendritic cells, macrophages and B cells. B cells activated by protein antigen attract and attach to T_H cells, to which they present the antigen via MHC II. If a T_H cell's receptors identify the antigen, the cell proliferates and releases cytokines which strongly potentiate B cell proliferation and performance; this is often essential for an effective response. T_H cytokines also induce B cell **class switching**, e.g. from production of IgM to IgE (Fig. 10b).

Memory cells which persist for years are also produced during clonal expansion. These respond much more rapidly and powerfully to subsequent exposures to the same pathogen, and provide **long term immunity**. This is the basis of immunization.

Cell-mediated immunity (Fig. 10d) is directed towards antigens *within* cells, which are made visible by MHC. **MHC I** is found on the surface of all cells and displays *cytosolic* antigens (e.g. viral proteins), but only to **cytotoxic T_C** (CD8+) cells, which proliferate on recognising the antigen and destroy any similarly infected cells. In contrast, MHC II displays antigens *retained within vesicles*, i.e. that have been phagocytosed, and is found *only* in APCs which activate T_H cells. **Dendritic cells**, and to a lesser extent macrophages, are the most important phagocytic APCs. After phagocytosing a pathogen they migrate to lymphoid tissues and present the antigen (via MHC II) to naïve T_H cells. T_H cells that recognise the antigen activate, proliferate and stimulate B cells (and thus a *humoral* response) as described above. Importantly, they also release cytokines that regulate the activity of other immune cells, including macrophages, T_H, T_C, NK, plasma and mast cells. T_H cells therefore play a critical coordinating role in the immune response.

11 Principles of diffusion and flow

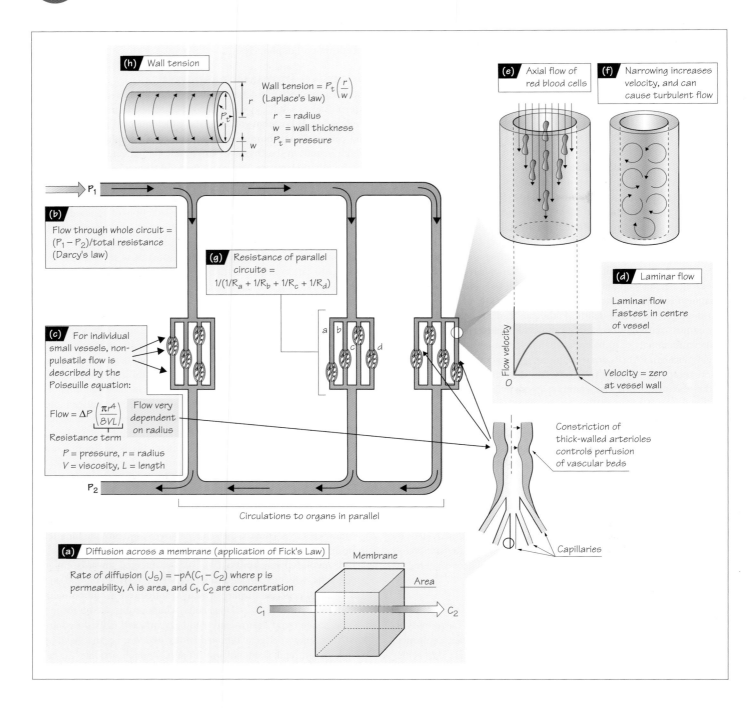

(h) Wall tension

$$\text{Wall tension} = P_t\left(\frac{r}{w}\right)$$
(Laplace's law)

r = radius
w = wall thickness
P_t = pressure

(e) Axial flow of red blood cells

(f) Narrowing increases velocity, and can cause turbulent flow

P_1

(b) Flow through whole circuit = $(P_1 - P_2)$/total resistance (Darcy's law)

(g) Resistance of parallel circuits =
$$1/(1/R_a + 1/R_b + 1/R_c + 1/R_d)$$

(d) Laminar flow

Laminar flow
Fastest in centre of vessel

Velocity = zero at vessel wall

Flow velocity

O

(c) For individual small vessels, non-pulsatile flow is described by the Poiseuille equation:

$$\text{Flow} = \Delta P\left(\frac{\pi r^4}{8VL}\right)$$
Resistance term

Flow very dependent on radius

P = pressure, r = radius
V = viscosity, L = length

P_2

Circulations to organs in parallel

Constriction of thick-walled arterioles controls perfusion of vascular beds

Capillaries

(a) Diffusion across a membrane (application of Fick's Law)

Rate of diffusion $(J_S) = -pA(C_1 - C_2)$ where p is permeability, A is area, and C_1, C_2 are concentration

Membrane

Area

C_1 ⟶ C_2

Physiology at a Glance, Third Edition. Jeremy P.T. Ward and Roger W.A. Linden.

 © 2013 John Wiley & Sons, Ltd. Published 2013 by John Wiley & Sons, Ltd.

Diffusion and bulk flow

Materials are carried around the body by a combination of bulk flow and diffusion. **Bulk flow** simply means transport *with* the carrying medium (blood, air). **Passive diffusion** refers to movement down a *concentration gradient*, and accounts for transport across small distances, e.g. within the cytosol and across membranes. The rate of **diffusion in a solution** is described by **Fick's law**:

$$J_s = -DA(\Delta C/\Delta x) \tag{11.1}$$

where J_s is the amount of substance transferred per unit time, ΔC is the difference in concentration, Δx is the diffusion distance and A is the surface area over which diffusion occurs. The negative sign reflects movement *down* the concentration gradient. D is the **diffusion coefficient**, a measure of how easy it is for the substance to diffuse. D is related to temperature, solvent viscosity and the size of the molecule, and is normally *inversely proportional to the cube root of the molecular weight*. **Diffusion across a membrane** is affected by the **permeability** of the membrane. The permeability (p) is related to the membrane thickness and composition, and the diffusion coefficient of the substance. Fick's equation can be rewritten as:

$$J_s = -pA\Delta C \tag{11.2}$$

where A is the membrane area and ΔC is the concentration difference across the membrane. The rate of diffusion across a capillary wall is therefore related to the *concentration difference* across the wall and the *permeability* of the wall to that substance (Fig. 11a).

Flow through a tube

Flow through a tube is dependent on the pressure difference across the ends of the tube ($P_1 - P_2$) and the resistance to flow provided by the tube (R):

$$Flow = (P_1 - P_2)/R \tag{11.3}$$

This is **Darcy's law** (analogous to *Ohm's law* in electronics; Fig. 11b).

Resistance is due to frictional forces, and is determined by the **diameter** of the tube and the **viscosity** of the fluid:

$$R = (8VL)/(\pi r^4) \tag{11.4}$$

This is **Poiseuille's law**, where V is the viscosity, L is the length of the tube and r is the radius of the tube. Combining eqns. (11.3) and (11.4) shows an important principle, namely that **flow ∝ (radius)^4**:

$$Flow = [(P_1 - P_2)\pi r^4]/(8VL) \tag{11.5}$$

Therefore, small changes in radius have a large effect on flow (Fig. 11c). Thus, the constriction of an artery by 20% will decrease the flow by ~60%.

Viscosity. Treacle flows more slowly than water because it has a higher viscosity. Plasma has a similar viscosity to water, but blood contains cells (mostly erythrocytes) which effectively increase the viscosity by three- to four-fold. Changes in cell number, e.g. *poly-cythaemia* (increased erythrocytes), therefore affect the blood flow.

Laminar and turbulent flow. Frictional forces at the sides of a tube cause drag on the fluid touching them. This creates a *velocity gradient* (Fig. 11d) in which the flow is greatest at the centre. This is termed **laminar flow**, and describes the flow in the majority of cardiovascular and respiratory systems at rest. A consequence of the velocity gradient is that blood cells tend to move away from the sides of the vessel and accumulate towards the centre (**axial streaming**; Fig.11e); they also

tend to align themselves to the flow. In small vessels, this *effectively reduces the blood viscosity* and minimizes the resistance (the **Fåhraeus–Lindqvist effect**).

At high velocities, especially in large arteries and airways, and at the edges or branches where the velocity increases sharply, flow may become **turbulent**, and laminar flow is disrupted (Fig. 11f). This significantly increases the resistance. The narrowing of airways and large arteries (or valve orifices), which increases the fluid velocity, can therefore cause turbulence, which is heard as **lung sounds** (e.g. wheezing in asthma) and **cardiac murmurs** (Chapter 18).

Turbulent flow is also responsible for the sounds heard when measuring blood pressure using a **sphygmomanometer** and stethoscope (**Korotkoff sounds**). A rubber cuff round the arm is inflated to a pressure well above predicted arterial pressure and then slowly deflated. When the pressure in the cuff approaches systolic pressure, the blood is able to force its way through the constricted artery in the arm for part of the pulse. The high velocity of the blood through the narrowed artery causes turbulence and therefore a sound; the first appearance of this is taken as systolic pressure. As the pressure in the cuff falls further and so below diastolic pressure, flow is continuous because the pressure is greater than that in the cuff throughout the pulse. As a result the sound fades and disappears, and the cuff pressure at this point is taken as diastolic pressure.

Resistances in parallel and in series. The cardiovascular and respiratory systems contain a mixture of *series* (e.g. arteries ⇒ arterioles ⇒ capillaries ⇒ venules ⇒ veins) and *parallel* (e.g. lots of capillaries) components (Fig. 11g). Flow through a *series* of tubes is restricted by the resistance of each tube in turn, and the total resistance is the **sum** of the resistances:

$$R_T = R_1 + R_2 + R_3 + \dots \tag{11.6}$$

In a parallel circuit, the addition of extra paths reduces the total resistance, and so:

$$R_T = 1/(1/R_a + 1/R_b \dots) \tag{11.7}$$

Although the resistance of individual capillaries or terminal bronchioles is high (small radius, *Poiseuille's law*), the huge number of them in parallel means that their contribution to the total resistance of the cardiovascular and respiratory systems is comparatively small.

Wall tension and pressure in spherical or cylindrical containers

Pressure across the wall of a flexible tube (*transmural pressure*) tends to extend it, and increases wall tension. This can be described by **Laplace's law**:

$$P_t = (Tw)/r \tag{11.8}$$

where P_t is the transmural pressure, T is the wall tension, w is the wall thickness and r is the radius (Fig. 11h). Thus, a small bubble with the same wall tension as a larger bubble will contain a greater pressure, and will collapse into the larger bubble if they are joined. In the lung, small alveoli would collapse into larger ones were it not for *surfactant* which reduces the surface tension more strongly as the size of the alveolus decreases (Chapter 26). Laplace's law also means that a large dilated heart (e.g. heart failure) has to develop more wall tension (contractile force) in order to obtain the same ventricular pressure, making it less efficient.

12 Skeletal muscle and its contraction

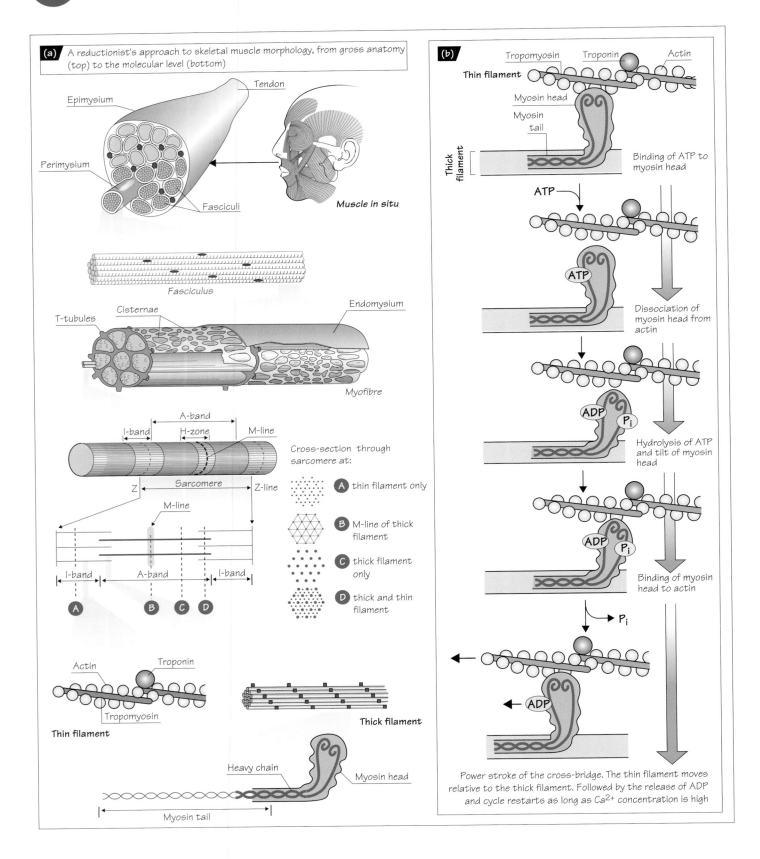

(a) A reductionist's approach to skeletal muscle morphology, from gross anatomy (top) to the molecular level (bottom)

Epimysium
Tendon
Perimysium
Fasciculi
Muscle in situ

Fasciculus

T-tubules
Cisternae
Endomysium
Myofibre

A-band
I-band
H-zone
M-line
Cross-section through sarcomere at:
Z
Sarcomere
Z-line

M-line

I-band
A-band
I-band
A B C D

A thin filament only

B M-line of thick filament

C thick filament only

D thick and thin filament

Actin
Troponin
Tropomyosin
Thin filament

Thick filament

Heavy chain
Myosin head
Myosin tail

(b)

Tropomyosin
Troponin
Actin
Thin filament
Myosin head
Myosin tail
Thick filament

Binding of ATP to myosin head

ATP

ATP

Dissociation of myosin head from actin

ADP
Pi

Hydrolysis of ATP and tilt of myosin head

ADP
Pi

Binding of myosin head to actin

Pi

ADP

Power stroke of the cross-bridge. The thin filament moves relative to the thick filament. Followed by the release of ADP and cycle restarts as long as Ca^{2+} concentration is high

Physiology at a Glance, Third Edition. Jeremy P.T. Ward and Roger W.A. Linden.

 © 2013 John Wiley & Sons, Ltd. Published 2013 by John Wiley & Sons, Ltd.

Muscles make up about 50% of the adult body mass. There are three types of muscle: **skeletal** (muscle attached to the skeleton); **cardiac** (muscle involved in cardiac function; Chapter 15) (both of these are morphologically striated or striped and are commonly called **striated muscles**); and **smooth** (muscle involved in many involuntary processes in the blood vessels and gut; this type is not structurally striated, hence its name; Chapter 15). A comparison of the properties of the three muscle types is shown in Appendix I.

Skeletal muscle

The **skeletal muscles** and the **skeleton** function together as the musculoskeletal system. Skeletal muscle is sometimes referred to as **voluntary muscle** because it is under conscious control. It uses about 25% of our oxygen consumption at rest and this can increase up to 20-fold during exercise.

General mechanisms of skeletal muscle contraction

The functions of muscle tissue are the development of tension and shortening of the muscle. Muscle fibres have the ability to shorten a considerable amount, which is brought about by the molecules sliding over each other. Muscle activity is transferred to the skeleton by the tendons, and the tension developed by the muscles is graded and adjusted to the load.

Fine structure of skeletal muscle (Fig. 12a)

The connective tissue surrounding the whole muscle is called the **epimysium**. The connective tissue that extends beyond the body of the muscle eventually blends into a **tendon**, which is attached to bone or cartilage. Skeletal muscle is composed of numerous parallel, elongated, multinucleated (up to 100) cells, referred to as **muscle fibres** or **myofibres**, which are between 10 and 100 μm in diameter and vary in length, and are grouped together to form **fasciculi**. Each fasciculus is surrounded by the **perimysium**. Each myofibre is encased by connective tissue called the **endomysium**. Beneath the endomysium is the **sarcolemma** (an excitable membrane). This is an elastic sheath with infoldings that invaginate the fibre interior, particularly at the motor end plate of the neuromuscular junction (Chapter 13). Each myofibre is made up of **myofibrils** 1 μm in diameter separated by cytoplasm and arranged in a parallel fashion along the long axis of the cell. Each myofibril is further subdivided into **thick** and **thin myofilaments** (thick, 10–14 nm in width and 1.6 μm in length; thin, 7 nm in width and 1 μm in length). These are responsible for the cross-striations. **Thin filaments** consist primarily of three proteins, **actin**, **tropomyosin** and **troponin**, in the ratio 7 : 1 : 1, and **thick filaments** consist primarily of **myosin**. The cytoplasm surrounding the **myofilaments** is called the **sarcoplasm**. Each myofibre is divided at regular intervals along its length into **sarcomeres** separated by **Z-discs** (in longitudinal sections, these are **Z-lines**). To the Z-lines are attached the thin filaments held in a hexagonal array. The **I-band** extends from either side of the Z-line to the beginning of the thick filament (myosin). The myosin filaments make up the **A-band**.

The **H-zone** is at the centre of the sarcomere, and the **M-line** is a disc of delicate filaments in the middle of the H-zone that holds the myosin filaments in position so that each one is surrounded by six actin filaments.

The thin filaments consist of two intertwining strands of actin with smaller strands of tropomyosin and troponin between the intertwining strands. Each strand of actin is made up of about 200 units of globular or G-actin. It is on these globules that there is a site for myosin to bind during contraction.

The thick filaments are made up of about 100 myosin molecules; each molecule is club shaped, with a thin tail (shaft) comprising two coiled peptide chains and a head made up of two heavy peptide chains and four light peptide chains that have a regulatory function. The ATPase activity of the myosin molecule is concentrated in the head.

The thin tails of the myosin molecules form the bulk of the thick filaments, whereas the heads are 'hinged' and project outward to form cross-bridges between the thick filaments and their neighbouring thin filaments. Six thin filaments surround each thick filament.

Between the myofibrils are a large number of mitochondria and glycogen granules, as found in other cells, but muscle cells have regular invaginations which project from outside the cell and wrap around the sarcomeres, particularly where the thin and thick filaments overlap. These invaginations are called transverse or **T-tubules** and contain extracellular fluid. The specialized smooth endoplasmic reticulum, the **sarcoplasmic reticulum**, which is close to the T-tubules, is enlarged to form **terminal cisternae** which actively transport Ca^{2+} into the lumen from the sarcoplasm.

Like fingers of the hands sliding over one another, actin and myosin molecules slide past each other. The myosin heads bind to the actin chain and tilt. There is a constant process of binding, tilting, releasing and rebinding of cross-bridges, as well as rotation of the myosin filaments as they interact with the actin filaments and bind with the alternate myofibril in the hexagonal structure. This results in the contraction of the whole muscle. The cross-bridges are formed asynchronously so that some are active, whilst others are resting.

The interaction of actin (thin filaments) and myosin (thick filaments) brings about contraction of the muscle, which is caused by the cross-bridges, a result of the interaction of troponin and Ca^{2+}. This mechanism is called the **sliding filament theory**. The contraction of muscle is triggered by the release of Ca^{2+} from the sarcoplasmic reticulum. Ca^{2+} floods out of the cisternae, where it is stored by binding reversibly with a protein, **calsequestrin**. This raises the concentration of calcium from 0.1 μmol/L to more than 10 μmol/L[1], saturating the binding sites on troponin. This results in a shift of tropomyosin, thus allowing the myosin cross-bridges to bind more strongly with actin and begin the contraction cycle (Fig. 12b). The heads tilt after attachment by hydrolysing the adenosine triphosphate (ATP) energy stores, releasing adenosine diphosphate (ADP) and inorganic phosphate (P_i), which leads to a greater binding of the cross-bridges. ADP and P_i escape from the head, freeing the head for another molecule of ATP. This releases the binding of the head and, if Ca^{2+} is still present, the cycle continues. Otherwise, the binding is inhibited. Contraction is maintained as long as Ca^{2+} is high. The duration of the contraction is dependent on the rate at which the sarcoplasmic reticulum pumps back the Ca^{2+} into the terminal cisternae.

13 Neuromuscular junction and whole muscle contraction

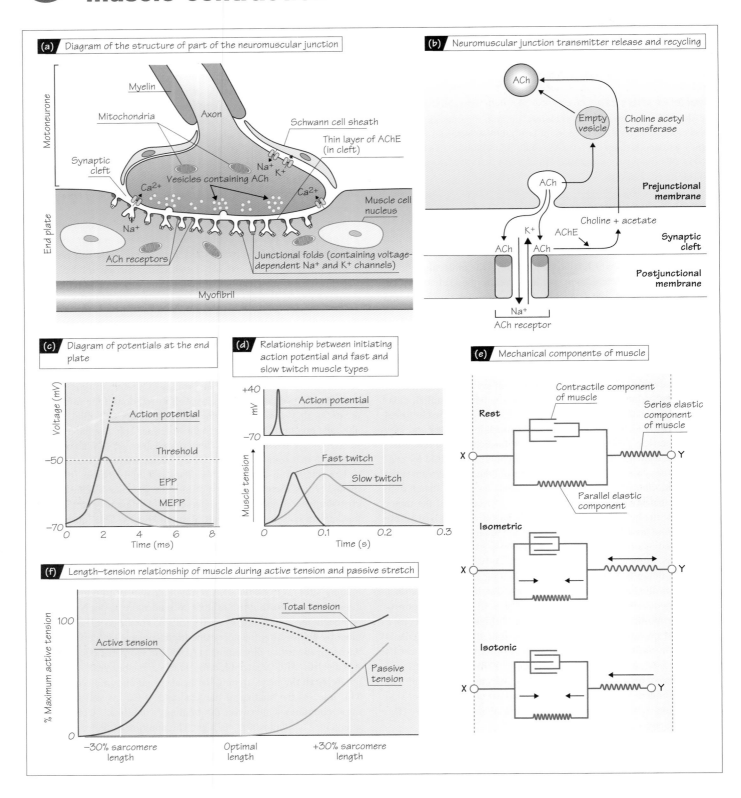

(a) Diagram of the structure of part of the neuromuscular junction

Motoneurone

Myelin

Mitochondria

Axon

Schwann cell sheath

Thin layer of AChE (in cleft)

Synaptic cleft

Na^+ K^+

Ca^{2+}

Vesicles containing ACh

Ca^{2+}

Muscle cell nucleus

End plate

Na^+

ACh receptors

Junctional folds (containing voltage-dependent Na^+ and K^+ channels)

Myofibril

(b) Neuromuscular junction transmitter release and recycling

ACh

Empty vesicle

Choline acetyl transferase

ACh

Prejunctional membrane

K+

AChE

Choline + acetate

ACh

ACh

Synaptic cleft

Postjunctional membrane

Na^+

ACh receptor

(c) Diagram of potentials at the end plate

Voltage (mV)

Action potential

Threshold

−50

EPP

MEPP

−70

0   2   4   6   8

Time (ms)

(d) Relationship between initiating action potential and fast and slow twitch muscle types

+40

mV

Action potential

−70

Muscle tension

Fast twitch

Slow twitch

0   0.1   0.2   0.3

Time (s)

(e) Mechanical components of muscle

Rest

Contractile component of muscle

Series elastic component of muscle

X   Y

Parallel elastic component

Isometric

X   Y

Isotonic

X   Y

(f) Length–tension relationship of muscle during active tension and passive stretch

% Maximum active tension

100

Total tension

Active tension

Passive tension

0

−30% sarcomere length

Optimal length

+30% sarcomere length

Physiology at a Glance, Third Edition. Jeremy P.T. Ward and Roger W.A. Linden.

© 2013 John Wiley & Sons, Ltd. Published 2013 by John Wiley & Sons, Ltd.

Neuromuscular junction

For **skeletal** (**voluntary**) **muscle** to contract, there must be neuronal activation to the muscle fibres themselves from either higher centres in the brain or via reflex pathways involving either the spinal cord or the brain stem. The neurones that innervate skeletal muscles are called **α-motor neurones**. Each motor axon splits into a number of branches that make contact with the surface of individual muscle fibres in the form of bulb-shaped endings. These endings make connections with a specialized structure on the surface of the muscle fibre, called the **motor end plate**, and together form the **neuromuscular junction** (**NMJ**) (Fig. 13a).

The role of the NMJ is the one-to-one transmission of excitatory impulses from the α-motor neurone to the muscle fibres it innervates. It allows a reliable transmission of the impulses from nerve to muscle and produces a predictable response in the muscle. In other words, an action potential in the motor neurone must produce an action potential in the muscle fibres it innervates; this, in turn, must produce a contraction of the muscle fibres. The process by which the **NMJ** produces this one-to-one response is shown in Figures 13a and b.

The motor neurone axon terminal has a large number of vesicles containing the transmitter substance **acetylcholine** (**ACh**). At rest, when not stimulated, a small number of these vesicles release their contents, by a process called **exocytosis**, into the synaptic cleft between the neurones and the muscle fibres. ACh diffuses across the cleft and reacts with specific ACh receptor proteins in the postsynaptic membrane (motor end plate). These receptors contain an integral ion channel, which opens and allows the movement inwards (influx) of small cations, mainly Na^+. There are more than 10^7 receptors on each end plate (**postjunctional membrane**); each of these can open for about 1 ms and allow small positively charged ions to enter the cell. This movement of positively charged ions generates an **end plate potential** (**EPP**). This is a depolarization of the cell with a rise-time of approximately 1–2 ms and may vary in amplitude (unlike the all-or-nothing response seen in the action potential; Chapter 6). The random release of ACh from the vesicles at rest gives rise to small, 0–4-mV depolarizations of the end plate, called **miniature end plate potentials** (**MEPP**) (Fig. 13c).

However, when an **action potential** reaches the **prejunctional nerve terminal**, there is an enhanced permeability of the membrane to Ca^{2+} ions due to opening of voltage-gated Ca^{2+} channels. This causes an increase in the exocytotic release of ACh from several hundred vesicles at the same time. This sudden volume of ACh diffuses across the cleft and stimulates a large number of receptors on the postsynaptic membrane, and thus produces an EPP that is above the threshold for triggering an action potential in the muscle fibre. It triggers a **self-propagating muscle action potential**. The depolarizing current (the **generator potential**) generated by the numbers of quanta of ACh is more than sufficient to cause the initiation of an action potential in the muscle membrane surrounding the postsynaptic junction. The typical summed **EPP** is usually four times the potential necessary to trigger an action potential in the muscle fibre, and so there is a large inherent safety factor.

The effect of ACh is rapidly abolished by the activity of the enzyme **acetylcholinesterase** (**AChE**). ACh is hydrolysed to **choline** and **acetic acid**. About one-half of the choline is recaptured by the presynaptic nerve terminal and used to make more ACh. Some ACh diffuses out of the cleft, but the enzyme destroys most of it. The number of vesicles available in the nerve ending is said to be sufficient for only about 2000 nerve–muscle impulses, and therefore the vesicles reform very rapidly within about 30 s (Fig. 13b).

Whole muscle contraction

As the action potential spreads over the muscle fibre, it invades the T-tubules and releases Ca^{2+} from the **sarcoplasmic reticulum** into the **sarcoplasm**, and the muscle fibres that are excited contract. This contraction will be maintained as long as the levels of Ca^{2+} are high. The single contraction of a muscle due to a single action potential is called a **muscle twitch**. Fibres are divided into **fast** and **slow twitch fibres** depending on the time course of their twitch contraction. This is determined by the type of myosin in the muscles and the amount of sarcoplasmic reticulum. Different muscles are made up of different proportions of these two types of fibre, leading to a huge variation in overall muscle contraction times (Fig. 3d).

At rest, muscles exert tension when stretched. Muscles have a passive elastic property and act both in series and parallel to the contractile element (Fig. 13e).

Isometric contraction occurs when the two ends of a muscle are held at a fixed distance apart, and stimulation of the muscle causes the development of tension within the muscle without a change in muscle length. **Isotonic contraction** occurs when one end of the muscle is free to move and the muscle shortens whilst exerting a constant force. In practice, most contractions are made up of both elements.

The relationships between resting, active and total tensions developed in skeletal muscle are shown in Figure 13f. The **passive curve** is due to the stretching of the elastic components, the **active curve** is due to contraction of the sarcomeres alone (contractile component), and the **total curve** is due to the sum of the passive and active tensions developed. It can be seen that the active tension developed is dependent on the length of the muscle. The optimum length occurs where the thick and thin filaments are thought to provide a maximum number of active cross-bridge sites for interaction (this length is very close to that of the resting length of a particular muscle). As the muscle length is increased, the thick and thin filaments overlap less, providing fewer cross-bridge sites for interaction; as the muscle shortens below the optimal length, the thin filaments overlap one another and, in so doing, reduce the number of active sites available for interaction with the thick filaments.

14 Motor units, recruitment and summation

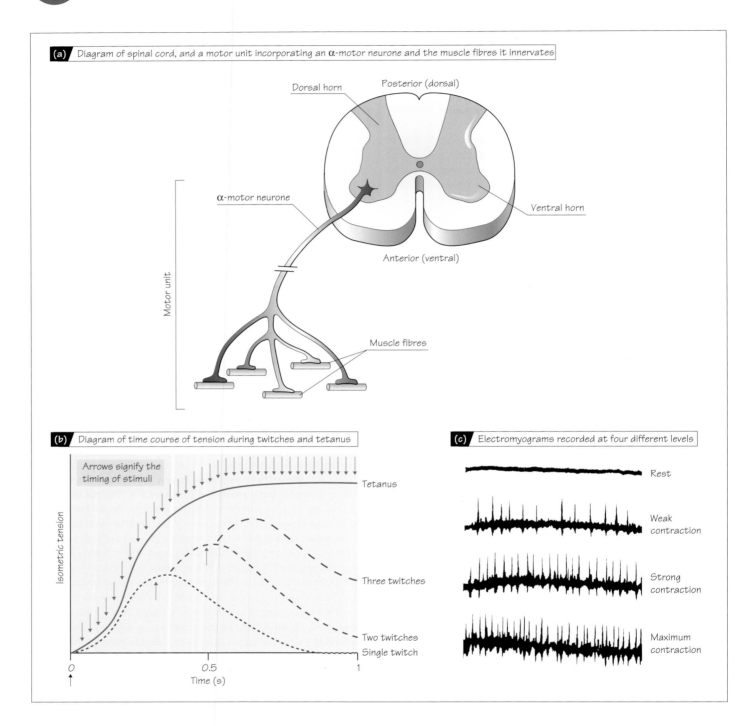

(a) Diagram of spinal cord, and a motor unit incorporating an α-motor neurone and the muscle fibres it innervates

Dorsal horn

Posterior (dorsal)

α-motor neurone

Ventral horn

Anterior (ventral)

Motor unit

Muscle fibres

(b) Diagram of time course of tension during twitches and tetanus

Arrows signify the timing of stimuli

Isometric tension

Tetanus

Three twitches

Two twitches
Single twitch

0   0.5   1
Time (s)

(c) Electromyograms recorded at four different levels

Rest

Weak contraction

Strong contraction

Maximum contraction

Physiology at a Glance, Third Edition. Jeremy P.T. Ward and Roger W.A. Linden.

© 2013 John Wiley & Sons, Ltd. Published 2013 by John Wiley & Sons, Ltd.

In normal skeletal muscle, fibres never contract as isolated individuals. Several contract at almost the same time, as they are all supplied by the same **α-motor neurone**. The single motor neurone and all the fibres it innervates is called the **motor unit** (Fig. 14a). This is the smallest part of a muscle that can be made to contract independently of other parts of the muscle. The number of muscle fibres innervated by one motor unit can be as low as 5 or as high as 2000. The number is correlated with the precision with which the tension developed by the muscle is graded.

Within the muscle, the muscle fibres of each motor unit are widely distributed amongst the fibres of many other units. This, in effect, distributes the demands made on the muscle's circulatory support. The ratio between the number of α-motor neurones and the total number of skeletal muscle fibres is small in muscles such as the extraocular muscles that provide fine smooth movements (1:5), but large in muscles such as the gluteus maximus that need to generate powerful but coarse movements (1:>1000).

Fibres have been classified into three types, on the basis of various structural and functional properties of motor units and their integral muscles. Table 14 outlines the relationship between the properties of the motor units (with their defining characteristics of conduction speed, resistance to fatigue and also size of activity pattern) and the properties of the muscle fibres they contain (the type of myosin which determines the speed of contraction, and the type of metabolism, which is highly correlated with the resistance to fatigue), and also the names given to these types in human muscles. Most muscles contain all three types (I, IIA, IIB), but differ in the proportions of each according to the function of the muscle as a whole. Posture muscles, such as the soleus, have mostly slow, fatigue-resistant, oxidative-type units, whereas movement muscles, such as the gastrocnemius, have a high proportion of the other two types. Training and exercising can alter these proportions.

The cell bodies of α-motor neurones also vary in size according to the type of motor unit: **motor neurones innervating type I fibres have the smallest cell bodies, and those innervating type IIB fibres have the largest**.

During graded contraction, there is a recruitment order of the units, such that the smallest cells discharge first and the largest last (the so-called **size principle**). **Force** is controlled not only by **varying the unit recruitment**, but also by **varying the firing rate of the units**. A single action potential in a single motor unit produces a delayed rise in tension in all the muscle fibres that make up that motor unit. A second and third action potential that occur soon after the first produce a summed contraction, or a series of twitches. The tension developed by the first action potential has not completely decayed when the second contraction is grafted on to the first, and so on for the third action potential and contraction. This is called **summation** (Fig. 14b). If the muscle fibres are stimulated repeatedly at a faster frequency, a sustained contraction results in which individual twitches cannot be detected. This is called **tetanus**. The tension of tetanus is much greater than the maximum tension of a single, double or triple twitch (Fig. 14b). For most units, the firing rate for a steady contraction is between 5 and 8 Hz. It can rise to 40 Hz or more, but only for very brief periods. During a gradual increase in contraction of a muscle, the first units start to discharge and increase their firing rate and, as the force needs to increase, new units are recruited and, in turn, also increase their firing rate. When there is a need to gradually decrease the force output, the pattern is reversed, so that those units that were recruited last will be the first to decrease their firing and then stop, and the last units to fire will be the smallest units. Because the unitary firing rates for each motor unit are different and not synchronized, the overall effect is a smooth force profile from the muscle. The greater the desynchronized firing, the smoother the movements observed. When synchronized firing does occur, such as in fatigued states and Parkinson's disease, marked muscle tremors are seen.

The summed excitatory impulses (action potentials) of the motor units can be recorded in an **electromyogram** (**EMG**). The EMG is an extracellular recording made from either the skin surface overlying a muscle or from electrodes inserted extracellularly within the body of the muscle. The increase in recruitment of individual motor units (**motor unit recruitment**), as well as the increased rate of firing of the units (**rate or frequency recruitment**), can sometimes be seen in the EMG during increased force of contraction (Fig. 14c).

Table 14 Properties of motor units and muscle fibres

Motor unit properties			
Contraction speed	Slow	Fast	Fast
Resistance to fatigue	Resistant	Resistant	Fatigable
Motor unit force	Small	Larger	Largest
Motor unit activity pattern	Long-lasting low frequency, e.g. postural	Frequent short bursts of moderate force	Rare, very short bursts for high forces
Motor unit name	SR (slow resistant)	FR (fast resistant)	FF (fast fatigable)
Muscle fibre properties			
Myosin isoform	I	IIA	IIX
Contraction speed	Slow	Fast	Fast
Metabolic process	Oxidative	Oxidative and glycolytic	Glycolytic
Capillary density	High	Medium	Low
Fibre type name			
Physiological	SO (slow oxidative)	FOG (fast oxidative and glycolytic)	FG (fast glycolytic)
Diagnostic	Type I	Type IIA	Type IIB*

*Type IIB will probably change to Type IIX to match myosin isoform in future.

15 Cardiac and smooth muscle

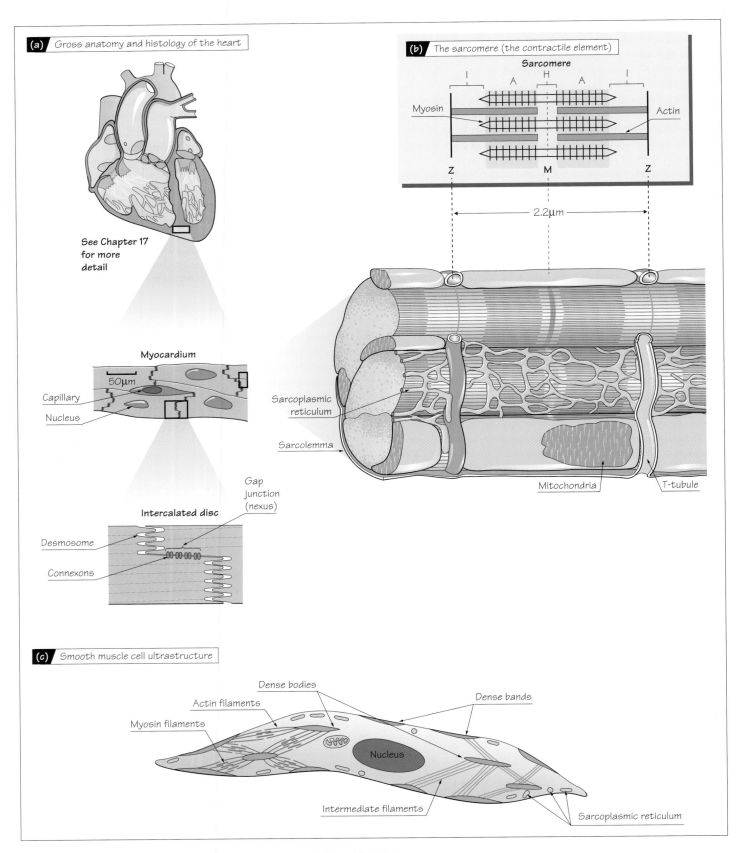

(a) Gross anatomy and histology of the heart

See Chapter 17 for more detail

Myocardium

50μm

Capillary

Nucleus

Gap junction (nexus)

Intercalated disc

Desmosome

Connexons

(b) The sarcomere (the contractile element)

Sarcomere

I A H A I

Myosin

Actin

Z M Z

2.2μm

Sarcoplasmic reticulum

Sarcolemma

Mitochondria

T-tubule

(c) Smooth muscle cell ultrastructure

Dense bodies

Actin filaments

Dense bands

Myosin filaments

Nucleus

Intermediate filaments

Sarcoplasmic reticulum

Physiology at a Glance, Third Edition. Jeremy P.T. Ward and Roger W.A. Linden.

 © 2013 John Wiley & Sons, Ltd. Published 2013 by John Wiley & Sons, Ltd.

Cardiac muscle

The muscles of the heart, the **myocardium**, generate the force of contraction of the atrial and ventricular muscles. The myocardium is composed of cardiac muscle cells called **myocytes**. These cells are striated due to the orderly arrangement of the thick and thin filaments which, as in skeletal muscle, make up the bulk of the muscle. However, they are less organized than in skeletal muscle (Fig. 15a,b). The **myocytes** have dimensions of $100 \times 20\,\mu m$, are branched, with a **single nucleus**, and are also **rich in mitochondria**. The normal pumping action of the heart is dependent on the synchronized contraction of all cardiac cells. Their contraction is not dependent on an external nerve supply, as in skeletal muscle, but instead the heart generates its own rhythm, called **inherent rhythmicity**. The nerves innervating the heart only speed up or slow down the rhythm and can modify the force of contraction (the so-called **chronotropic** and **inotropic** effects respectively; Chapter 19).

The synchronicity between myocytes occurs because all the adjacent cells are linked to one another at their ends by **specialized gap** or **electrotonic junctions** – intercalated discs – which are essentially low-resistance pathways between cells. These allow action potentials to spread rapidly from one cell to another and enable the cardiac muscle to act as a **functional syncytium** (i.e. it acts as a single unit although comprising individual cells).

The intercalated discs provide both a structural attachment (**desmosomes**) between cells and an electrical contact, called a gap junction, made up of proteins called **connexons** (Fig. 15a). Although a rise in intracellular Ca^{2+} initiates contraction in the same way as in skeletal muscle (Chapter 12), the mechanisms leading to this rise in intracellular Ca^{2+} are fundamentally different, and are discussed in Chapter 19.

Smooth muscle

The **absence of striations** within the cells and the poorer organization of the fibres give this type of muscle its name. Each cell contains only **one nucleus** situated near the centre. Smooth muscle is involved in many involuntary processes in blood vessels and the gut.

The smooth muscle of each organ is distinctive from that of most other organs, and there is considerable variation in the structure and function of smooth muscle in different parts of the body; however, essentially it can be divided into **unitary** (or **visceral**) and **multiunit** smooth muscle types.

Smooth muscle cells are spindle-shaped with dimensions of $50–400\,\mu m$ in length by $2–10\,\mu m$ thick. They are joined, like cardiac muscle, by special intercellular connections called **desmosomes**. Because the actin and myosin filaments are not regularly arranged, they lack striations. Smooth muscle cells shorten by sliding of the myofilaments towards and over one another, but at a much slower rate than in other muscle types. For this reason, they are capable of prolonged, maintained contraction, without fatigue and with little energy consumption (Fig. 15c).

The **unitary muscle type** or **visceral smooth muscle** exhibits many gap junctions between cells, and a steady wave of contraction can pass through a whole sheet of muscle as if it were a single unit. It is commonly found in the stomach, intestines, urinary bladder, urethra and blood vessels, and is capable of bringing about **autorhythmical activity** (seen particularly in the digestive tract where it is modulated by neuronal activity).

Tonic activity causes smooth muscle to remain in a constant state of contraction or tonus. It is commonly found in sphincters that control the movement of digestive products through the gastrointestinal tract.

Multiunit smooth muscle is made up of individual fibres not connected by gap junctions, but separately stimulated by autonomic motor neurones. Each smooth muscle fibre can contract independently from the others. Examples include the ciliary muscles of the eye, the iris of the eye and piloerector muscles that cause erection of hairs when stimulated by the sympathetic system.

The factors that influence the neural control of smooth muscle are:
1 The type of innervation and the transmitter released.
2 The receptor of the neurotransmitter on the muscle cell itself.
3 The anatomical arrangement of the nerve in relation to the muscle fibres.

There are three types of innervation: **extrinsic** – from the autonomic part of the nervous system, mainly sympathetic (arteries), parasympathetic (ciliary muscles) and both sympathetic and parasympathetic (gut); **intrinsic** – a plexus of nerves within the smooth muscle itself (seen in the gut); and **afferent sensory neurones** – these indirectly lead to the reflex activation of motor neurones.

Smooth muscle cells also respond to local tissue factors and hormones, i.e. changes in the fluids that surround them (interstitial fluids). In addition, many hormones that circulate in the bloodstream also cause smooth muscle contraction [hormones such as adrenaline (epinephrine), angiotensin, oxytocin, antidiuretic hormone (ADH), noradrenaline (norepinephrine) and serotonin]. Also, a lack of oxygen in the tissues causes smooth muscle cells to relax and vasodilate; an increase in CO_2 or H^+ also causes vasodilatation (Chapter 21).

Contractile mechanisms of smooth muscle

Smooth muscle contains no troponin, but has twice as much actin and tropomyosin as striated muscle. Myosin is also present, but only in about one-quarter of the amount found in striated muscle fibres. The rate at which cross-bridges are formed and released is slower (some 300 times) than that in striated muscle fibres, in part due to the different mechanisms involved.

Although contraction of smooth muscle is initiated by an increase in Ca^{2+}, unlike striated muscle this is not mediated via the interaction of Ca^{2+} with troponin (there is none). Instead, cross-bridge formation is controlled from the myosin side in a rather more complex fashion. Ca^{2+} binds to the protein **calmodulin**, which activates **myosin light chain kinase**. This phosphorylates myosin which allows it to form cross-bridges with actin, using energy from adenosine triphosphate (ATP). It follows that there must be a means by which myosin is dephosphorylated. This is provided by **myosin phosphatase**. Many factors that contract smooth muscle do so by inhibiting myosin phosphatase at the same time as raising Ca^{2+}, so maximising the degree of myosin phosphorylation (Chapter 21).

A comparison of the properties of skeletal, cardiac and smooth muscle is shown in the Appendix I.

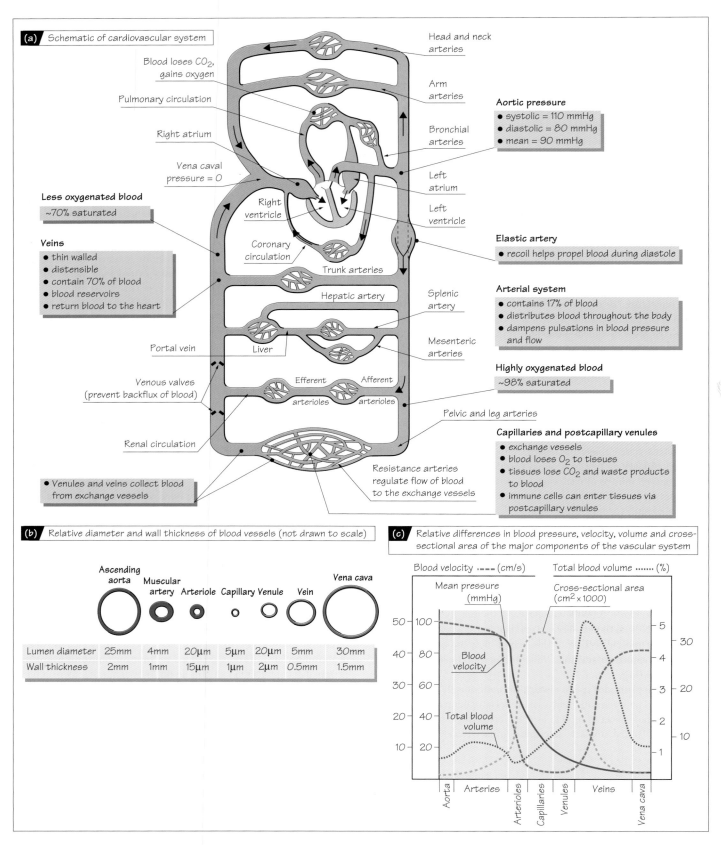

(a) Schematic of cardiovascular system

Head and neck arteries

Blood loses CO_2, gains oxygen

Pulmonary circulation

Arm arteries

Bronchial arteries

Aortic pressure
- systolic = 110 mmHg
- diastolic = 80 mmHg
- mean = 90 mmHg

Right atrium

Vena caval pressure = 0

Left atrium

Less oxygenated blood

~70% saturated

Right ventricle

Left ventricle

Veins
- thin walled
- distensible
- contain 70% of blood
- blood reservoirs
- return blood to the heart

Coronary circulation

Trunk arteries

Elastic artery
- recoil helps propel blood during diastole

Hepatic artery

Splenic artery

Arterial system
- contains 17% of blood
- distributes blood throughout the body
- dampens pulsations in blood pressure and flow

Portal vein

Liver

Mesenteric arteries

Highly oxygenated blood

~98% saturated

Venous valves (prevent backflux of blood)

Efferent arterioles

Afferent arterioles

Pelvic and leg arteries

Capillaries and postcapillary venules
- exchange vessels
- blood loses O_2 to tissues
- tissues lose CO_2 and waste products to blood
- immune cells can enter tissues via postcapillary venules

Renal circulation

- Venules and veins collect blood from exchange vessels

Resistance arteries regulate flow of blood to the exchange vessels

(b) Relative diameter and wall thickness of blood vessels (not drawn to scale)

Ascending aorta Muscular artery Arteriole Capillary Venule Vein Vena cava

	Ascending aorta	Muscular artery	Arteriole	Capillary	Venule	Vein	Vena cava
Lumen diameter	25mm	4mm	20µm	5µm	20µm	5mm	30mm
Wall thickness	2mm	1mm	15µm	1µm	2µm	0.5mm	1.5mm

(c) Relative differences in blood pressure, velocity, volume and cross-sectional area of the major components of the vascular system

Blood velocity ---- (cm/s)   Total blood volume .... (%)

Mean pressure (mmHg)

Cross-sectional area (cm^2 × 1000)

Blood velocity

Total blood volume

Aorta Arteries Arterioles Capillaries Venules Veins Vena cava

Physiology at a Glance, Third Edition. Jeremy P.T. Ward and Roger W.A. Linden.

 © 2013 John Wiley & Sons, Ltd. Published 2013 by John Wiley & Sons, Ltd.

The cardiovascular system comprises the heart and blood vessels, and contains ~5.5 L of blood in a 70 g man. Its main functions are to distribute O_2 and nutrients to tissues, transfer metabolites and CO_2 to excretory organs and the lungs, and transport hormones and components of the immune system. It is also important for thermoregulation. The cardiovascular system is arranged mostly in **parallel**, i.e. each tissue derives blood directly from the aorta (Fig. 16a). This allows all tissues to receive fully oxygenated blood, and flow can be controlled independently in each tissue against a constant pressure head by altering the resistance of small arteries (i.e. arteriolar constriction or dilatation). The right heart, lungs and left heart are arranged in **series**. **Portal systems** are also arranged in series, where blood is used to transport materials directly from one tissue to another, such as the **hepatic portal system** between digestive organs and the liver. The function of the cardiovascular system is modulated by the autonomic nervous system (Chapter 7).

Blood vessels

The vascular system consists of **arteries** and **arterioles** that take blood from the heart to the tissues, thin-walled **capillaries** that allow the diffusion of gases and metabolites, and **venules** and **veins** that return blood to the heart. The blood pressure, vessel diameter and wall thickness vary throughout the circulation (Fig. 16b,c). Varying amounts of **smooth muscle** are contained within the vessel walls, allowing them to constrict and alter their resistance to flow (Chapters 11 and 21). Capillaries contain no smooth muscle. The inner surface of all blood vessels is lined with a thin monolayer of **endothelial cells**, important for vascular function (Chapter 21). Large arteries are **elastic** and partially damp out oscillations in pressure produced by the pumping of the heart; stiff arteries (age, atherosclerosis) result in larger oscillations. Small arteries contain relatively more muscle and are responsible for controlling tissue blood flow. Veins have a larger diameter than equivalent arteries, and provide less resistance. They have thin, distensible walls and contain ~70% of the total blood volume (Fig. 16c). Large veins are known as **capacitance vessels** and act as a *blood volume reservoir*; when required, they can constrict and increase the effective blood volume (Chapter 21). Large veins in the limbs contain **one-way valves**, so that when muscle activity (e.g. walking) intermittently compresses these veins they act as a pump, and assist the return of blood to the heart (the **muscle pump**).

The heart

The **heart** is a four-chambered muscular pump which propels blood around the circulation. It has an intrinsic pacemaker and requires no nervous input to beat normally, although it is modulated by the **autonomic nervous system** (Chapter 7). The volume of blood pumped per minute (**cardiac output**) is ~5 L at rest in humans, although this can increase to above 20 L during exercise. The volume ejected per beat (**stroke volume**) is ~70 mL at rest. The **ventricles** perform the work of pumping; **atria** assist ventricular filling. Unidirectional flow through

the heart is maintained by **valves** between the chambers and outflow tracts. Contraction of the heart is called **systole** (pronounced *sis'-to-ley*) ; the period between each systole, when the heart refills with blood, is called **diastole** (*di-as'-to-ley*).

The systemic circulation

During systole, the pressure in the left ventricle increases to ~120 mmHg, and blood is ejected into the **aorta**. The rise in pressure stretches the elastic walls of the aorta and large arteries, and drives blood flow. **Systolic pressure** is the maximum arterial pressure during systole (~110 mmHg). During diastole, arterial blood flow is partly maintained by elastic recoil of the walls of large arteries. The minimum pressure reached before the next systole is the **diastolic pressure** (~80 mmHg). The difference between the systolic and diastolic pressures is the **pulse pressure**. Blood pressure is expressed as the systolic/diastolic arterial pressure, e.g. 110/80 mmHg. The **mean blood pressure** (mean arterial pressure, MAP) cannot be calculated by averaging these pressures, because for ~60% of the time the heart is in diastole. It is instead estimated as the *diastolic pressure plus one-third of the pulse pressure*, e.g. $80 + 1/3(110 - 80) \approx 90$ mmHg.

The **major arteries** divide repeatedly into smaller **muscular arteries**, the smallest of which (diameter <100 μm) are called **arterioles**. Tissue blood flow is regulated by the constriction of these small arteries, referred to as **resistance vessels**. The mean blood pressure at the start of the arterioles is ~65 mmHg. The arterioles divide into dense networks of **capillaries** in the tissues, and these rejoin into small and then larger **venules**, the smallest veins. Capillaries and small venules provide the exchange surface between blood and tissues, contain no smooth muscle and are called **exchange vessels**; some gas exchange also occurs across the walls of small arterioles. The pressure on the arterial side of capillaries is ~25 mmHg and, on the venous side, ~15 mmHg. Venules converge into veins and finally the **vena cava**. This returns the partially deoxygenated and CO_2-loaded blood to the right atrium. The pressure in the vena cava at the level of the heart is called the **central venous pressure** (CVP), and is close to 0 mmHg.

The pulmonary circulation

The right atrium helps to fill the right ventricle, which pumps blood into the **pulmonary artery** and lungs. The pulmonary circulation is shorter than the systemic, and has a lower resistance to flow. Less pressure is therefore required to drive blood through the lungs; the pulmonary artery pressure is ~20/15 mmHg. Gas exchange occurs in capillaries surrounding the alveoli (small air sacs) of the lungs. These rejoin to form pulmonary venules and veins, and oxygenated blood is returned through the pulmonary vein to the left atrium, and hence to the left ventricle. The metabolic requirements of the lungs are not met by the pulmonary circulation, but by the separate **bronchial circulation**, the venous outflow of which returns to the left side of the heart (Fig. 16a).

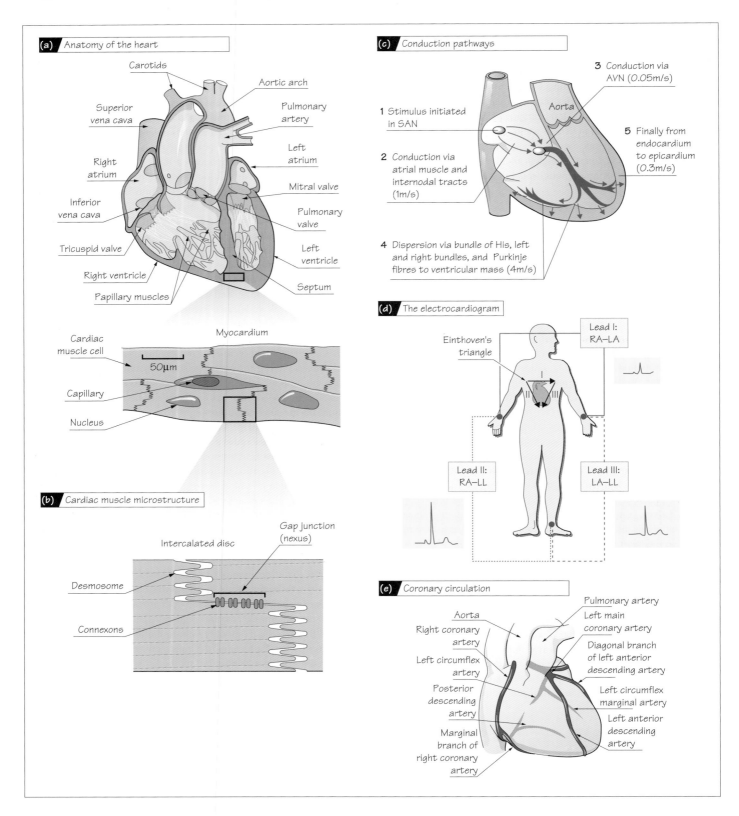

(a) Anatomy of the heart

Carotids
Aortic arch
Superior vena cava
Pulmonary artery
Right atrium
Left atrium
Inferior vena cava
Mitral valve
Tricuspid valve
Pulmonary valve
Right ventricle
Left ventricle
Papillary muscles
Septum

Myocardium

Cardiac muscle cell
50 μm
Capillary
Nucleus

(b) Cardiac muscle microstructure

Intercalated disc
Gap junction (nexus)
Desmosome
Connexons

(c) Conduction pathways

1 Stimulus initiated in SAN
2 Conduction via atrial muscle and internodal tracts (1m/s)
3 Conduction via AVN (0.05m/s)
Aorta
5 Finally from endocardium to epicardium (0.3m/s)
4 Dispersion via bundle of His, left and right bundles, and Purkinje fibres to ventricular mass (4m/s)

(d) The electrocardiogram

Einthoven's triangle
Lead I: RA–LA
Lead II: RA–LL
Lead III: LA–LL

(e) Coronary circulation

Aorta
Pulmonary artery
Right coronary artery
Left main coronary artery
Left circumflex artery
Diagonal branch of left anterior descending artery
Posterior descending artery
Left circumflex marginal artery
Marginal branch of right coronary artery
Left anterior descending artery

Physiology at a Glance, Third Edition. Jeremy P.T. Ward and Roger W.A. Linden.
 © 2013 John Wiley & Sons, Ltd. Published 2013 by John Wiley & Sons, Ltd.

The heart consists of four chambers – two thin-walled **atria** and two muscular **ventricles**. The atria are separated from the ventricles by a band of fibrous connective tissue (**annulus fibrosus**), which provides a skeleton for the attachment of muscle and the insertion of cardiac valves. It also prevents electrical conduction between the atria and ventricles, except at the **atrioventricular node** (AVN). The walls of the heart are formed from **cardiac muscle (myocardium)**. As the systemic circulation has a 10–15-fold greater resistance to flow than the pulmonary circulation, the left ventricle needs to develop more force and has more muscle than the right ventricle. The inner surface of the heart is covered by a thin layer of cells called the **endocardium**, similar to vascular **endothelial** cells (Chapter 21). This provides an anti-thrombogenic surface (inhibits clotting). The outer surface is covered by the **epicardium**, a layer of mesothelial cells. The whole heart is enclosed in a thin fibrous sheath (the **pericardium**), containing interstitial fluid as a lubricant, which protects the heart from damage caused by friction and prevents excessive enlargement.

Cardiac valves

Blood flows from the right atrium into the right ventricle via the **tricuspid** (three cusps or leaflets) atrioventricular (AV) valve, and from the left atrium to the left ventricle via the **mitral** (two cusps) AV valve. The AV valves are prevented from being everted into the atria by the high pressures developed in the ventricles by fine cords (**chordae tendinae** or *trabeculae*) attached between the edge of the valve cusps and **papillary muscles** in the ventricles (Fig. 17a). Blood is ejected from the right ventricle through the **pulmonary semilunar** valve into the pulmonary artery, and from the left ventricle via the **aortic semilunar** valve into the aorta; both semilunar valves have three cusps. The cusps are formed from connective tissue covered in a thin layer of **endocardial** or **endothelial** cells. When closed, the cusps form a tight seal at the **commissure** (line at which the edges meet). Both sets of valves open and close *passively* according to the pressure difference across them. Disease or the malformation of valves can have serious consequences. **Stenosis** describes narrowed valves; stenotic AV valves impair ventricular filling, and stenotic outflow valves increase **afterload** and thus ventricular work. **Incompetent** valves do not close properly and leak (**regurgitate**).

Cardiac pacemaker, conduction of the impulse and electrocardiogram

Cardiac muscle is described in Chapter 15. The heart beat is initiated in the **sinoatrial node** (SAN), a region of specialized myocytes in the right atrium, close to the coronary sinus. Spontaneous depolarization of the SAN (Chapter 19) provides the impulse for the heart to contract. Its rate is modulated by **autonomic nerves**. Action potentials (Chapter 19) in the SAN activate adjacent atrial myocytes via **gap junctions** contained within the **intercalated discs**; **desmosomes** provide a physical link (Fig. 17b; Chapter 19). A wave of depolarization and contrac-

tion therefore sweeps through the atrial muscle. This is prevented from reaching the ventricles directly by the **annulus fibrosus** (see above), and the impulse is channelled through the AVN, located between the right atrium and ventricle near the atrial septum.

The AVN contains small cells and conducts slowly; it therefore delays the impulse for ~120 ms, allowing time for atrial contraction to complete ventricular filling. Once complete, effective pumping requires rapid activation throughout the ventricles, and the impulse is transmitted from the AVN by specialized, wide and thus fast conducting myocytes in the **bundle of His** and **Purkinje fibres**, by which it is distributed over the inner surface of both ventricles (Fig. 17c). From here, a wave of depolarization and contraction moves from myocyte to myocyte across the endocardium until the whole ventricular mass is activated.

Electrocardiogram (Fig. 17d). The waves of depolarization through the heart cause *local currents* in surrounding fluid, which are detected at the body surface as small changes in voltage. This forms the basis of the **electrocardiogram** (ECG). The classical ECG records the voltage between the left and right arm (**lead I**), the right arm and left leg (**lead II**), and the left arm and left leg (**lead III**). This is represented by **Einthoven's triangle** (Fig. 17d). The size of voltage at any time depends on the quantity of muscle depolarizing (more cells generate more current), and the **direction** in which the wave of depolarization is travelling (i.e. it is a **vector** quantity). Thus, lead II normally shows the largest deflection during ventricular depolarization, as the muscle mass is greatest and depolarization travels from apex to base, more or less parallel to a line from the left hip to the right shoulder. The basic interpretation of the ECG is described in Chapter 18.

Coronary circulation

The heart requires a rich blood supply, which is derived from the **left** and **right coronary arteries** arising from the aortic sinus (Fig. 17e). Cardiac muscle has an extensive system of capillaries. Most of the blood returns to the right atrium via the **coronary sinus**. The **large** and **small** coronary veins run parallel to the right coronary arteries, and empty into the coronary sinus. Small vessels, such as the **thebesian veins**, empty into the cardiac chambers directly. The left ventricle is mostly supplied by the left coronary artery; occlusion in coronary artery disease can lead to serious damage. The coronary circulation is, however, capable of developing a good collateral system over time, where new arteries by-pass occlusions and improve perfusion. During systole, contraction of the ventricles compresses the coronary arteries and suppresses blood flow; this is of greatest effect in the left ventricle, where during systole the ventricular pressure is the same as or greater than that in the arteries. As a result, *more than 85% of left ventricular perfusion occurs during diastole*. This is problematic in disease if the heart rate is increased (e.g. exercise), as the diastolic interval is shorter.

18 The cardiac cycle

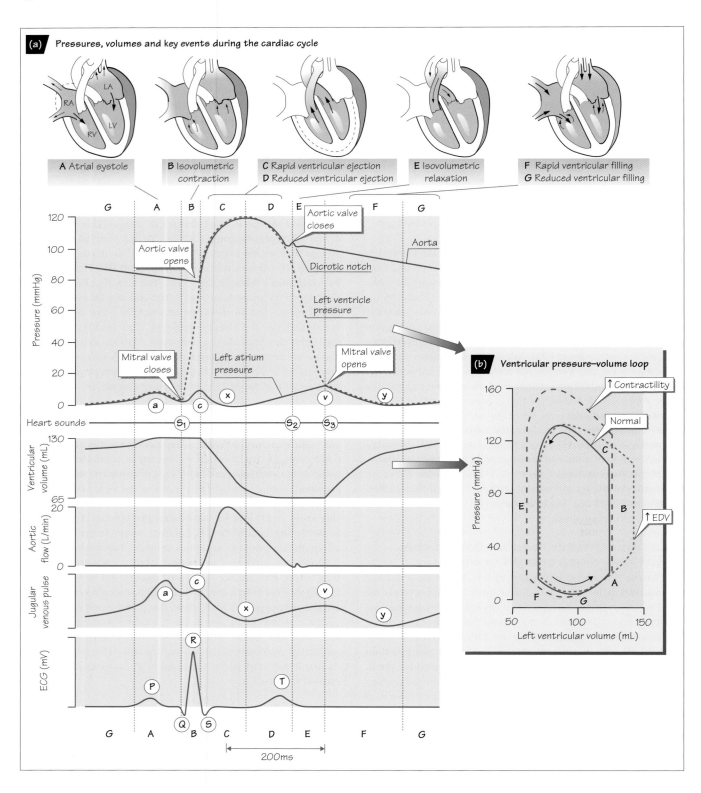

(a) Pressures, volumes and key events during the cardiac cycle

A Atrial systole

B Isovolumetric contraction

C Rapid ventricular ejection
D Reduced ventricular ejection

E Isovolumetric relaxation

F Rapid ventricular filling
G Reduced ventricular filling

Aortic valve closes

Aortic valve opens

Aorta

Dicrotic notch

Left ventricle pressure

Mitral valve closes

Left atrium pressure

Mitral valve opens

Heart sounds

Ventricular volume (mL)

Aortic flow (L/min)

Jugular venous pulse

ECG (mV)

200ms

(b) Ventricular pressure–volume loop

↑ Contractility

Normal

↑EDV

Pressure (mmHg)

Left ventricular volume (mL)

Physiology at a Glance, Third Edition. Jeremy P.T. Ward and Roger W.A. Linden.

© 2013 John Wiley & Sons, Ltd. Published 2013 by John Wiley & Sons, Ltd.

The cardiac cycle (Fig. 18a) describes the events that occur during one beat of the heart. These are shown in the figure for the left side of the heart, together with the pressures and volumes in the chambers and main vessels. At the start of the cycle, towards the end of **diastole**, the whole of the heart is relaxed. The atrioventricular (AV) valves are open (right, **tricuspid**; left, **mitral**), because the atrial pressure is still slightly greater than the ventricular pressure. The **pulmonary** and **aortic valves** are closed, as the pulmonary artery and aortic pressures are greater than that in the ventricles. The cycle starts when the **sinoatrial node** (SA node) initiates atrial systole (Chapter 19).

Atrial systole (A). At rest, atrial contraction only contributes the last ~15–20% of the final ventricular volume, as most of the filling has already occurred due to venous pressure. The atrial contribution increases with heart rate, as diastole shortens and there is less time for ventricular filling. There are no valves between the veins and atria, and atrial systole causes a small pressure rise in the great veins (**a wave**). The ventricular volume after filling is complete (**end-diastolic volume**, EDV) is ~120–140 mL in humans. The **end-diastolic pressure** (EDP) is less than 10 mmHg, and is higher in the left ventricle than in the right due to the thicker and therefore stiffer left ventricular wall. EDV strongly affects the strength of ventricular contraction (see **Starling's law**; Chapter 20). Atrial depolarization causes the **P wave** of the electrocardiogram (**ECG**); it should be noted that atrial *repolarization* is too diffuse to be seen on the ECG.

Ventricular systole (**B, C**). The ventricular pressure rises sharply during contraction, and the AV valves close as soon as this is greater than the atrial pressure. This causes a vibration which is heard as the **first heart sound** (S_1). Ventricular depolarization is associated with the **QRS** complex of the ECG. For a short period, while force is developing, both the AV and outflow (semilunar) valves are closed and there is no ejection, as the ventricular pressure is still less than that in the pulmonary artery and aorta. This is called **isovolumetric contraction** (**B**), as the ventricular volume does not change. The increasing pressure makes the AV valves bulge into the atria, causing a small atrial pressure wave (**c wave**), followed by a fall (**x descent**).

Ejection. Eventually, the ventricular pressure exceeds that in the aorta or pulmonary artery, the outflow valves open and blood is ejected. The flow is initially very rapid (**rapid ejection phase, C**) but, as contraction wanes, ejection is reduced (**reduced ejection phase, D**). During the second half of ejection, the ventricles stop actively contracting, and the muscle starts to repolarize; this is associated with the **T wave** of the ECG. The ventricular pressure during the reduced ejection phase is slightly less than that in the artery, but initially blood continues to flow out of the ventricle because of momentum. Eventually, the flow briefly reverses, causing the closure of the outflow valve, a small increase in aortic pressure (**dicrotic notch**) and the **second heart sound** (S_2). The amount of blood ejected in one beat is the **stroke volume**, ~70 mL. About 50 mL of blood is therefore left in the ventricle at the end of systole (**end-systolic volume, ESV**). The pro-portion of EDV that is ejected (stroke volume/EDV) is the **ejection fraction**; this is normally ~0.6, but is reduced below 0.5 in heart failure.

Diastole. Immediately after the closure of the outflow valves, the ventricles rapidly relax. The AV valves remain closed, however, because the ventricular pressure is initially still greater than that in the atria (**isovolumetric relaxation, E**). This is called isometric relaxation because again the ventricular volume does not change. Meanwhile, the atrial pressure has been increasing due to filling from the veins (**v wave**). When the ventricular pressure falls sufficiently, the AV valves open and the atrial pressure falls (**y descent**) as the ventricles rapidly refill (**rapid filling phase, F**). This is assisted by elastic recoil of the ventricular walls, essentially sucking blood into the ventricle. Filling during the last two-thirds of diastole is slower and due to venous flow alone (**reduced filling phase, G**). Diastole is twice the length of systole at rest, but decreases as the heart rate increases.

Ventricular pressure–volume loop

The ventricular pressure plotted against volume generates a loop (Fig. 18b), the area of which represents the work performed. Its shape is affected by the force of ventricular contraction (contractility), factors that alter refilling (**EDV**) and the pressure against which the ventricle has to pump (e.g. aortic pressure, **afterload**). An estimate of **stroke work** is calculated from the mean arterial pressure × stroke volume.

The pulse

The **peripheral arterial pulse** reflects the pressure waves travelling down through the blood from the heart; these move much faster than the blood itself. The shape of the pulse is affected by the compliance (stretchiness) and diameter of the artery; stiff (e.g. atherosclerosis) or small arteries have sharper pulses because they cannot absorb the energy so easily. Secondary peaks are due to reflections of the pressure wave at bifurcations of the artery. The **jugular venous pulse** reflects the right atrial pressure, as there is no valve between the jugular vein and right atria, and has corresponding **a**, **c**, and **v waves**.

Heart sounds

Heart sounds are caused by vibrations in the blood due, for example, to closure of the cardiac valves (see above). Normally, only the **first** and **second** heart sounds are detectable (S_1, S_2), although a third sound (S_3) can occasionally be heard in fit young people. When the atrial pressure is raised (e.g. in heart failure), both a third and fourth sound may be heard, associated with rapid filling and atrial systole, respectively; this sounds like a galloping horse (**gallop rhythm**). Cardiac **murmurs** are caused by turbulent blood, and a benign murmur is sometimes heard in young people during the ejection phase. Pathological murmurs are associated with the narrowing of valves (**stenosis**), or **regurgitation** of blood backwards through valves that do not close properly (**incompetence**).

19 Initiation of the heart beat and excitation–contraction coupling

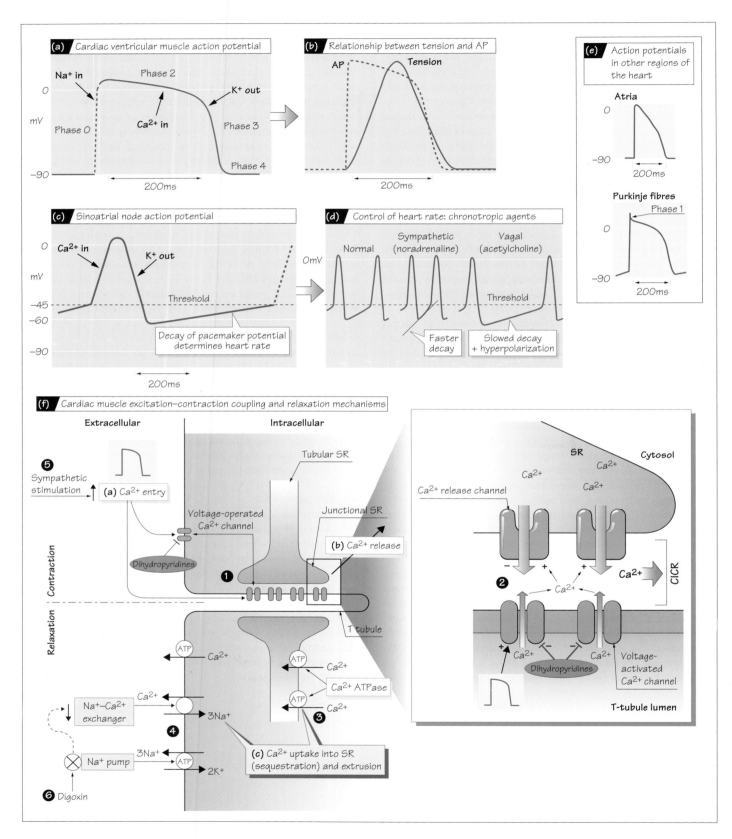

Physiology at a Glance, Third Edition. Jeremy P.T. Ward and Roger W.A. Linden.

The process linking depolarization to contraction is called **excitation–contraction coupling**. The basics of action potentials (**APs**) are described in Chapter 5.

Cardiac muscle electrophysiology

Ventricular muscle action potential (Fig. 19a). The resting potential of ventricular myocytes is approximately −90 mV (close to E_K) and stable (phase 4; Fig. 19a). An AP is initiated when the myocyte is depolarized to a **threshold potential** of approximately −65 mV, as a result of transmission from an adjacent myocyte via **gap junctions** (Chapter 17). Fast, voltage-gated Na^+ channels are activated, leading to an inward current which depolarizes the membrane rapidly towards +30 mV. This initial depolarization or **upstroke** (phase 0; Fig. 19a) is similar to that in nerve and skeletal muscle, and assists transmission to the next myocyte. The Na^+ current rapidly inactivates, but, in cardiac myocytes, the initial depolarization activates voltage-gated Ca^{2+} channels (**L-type channels**; threshold approximately −45 mV), through which Ca^{2+} floods into the cell. The resultant inward current prevents the cell from repolarizing, and causes a **plateau phase** (phase 2; Fig. 19a) that is maintained for ~250 ms until the L-type channels inactivate. The cardiac AP is thus much longer than that in nerve or skeletal muscle (~300 ms vs ~2 ms). Repolarization occurs due to activation of a voltage-gated outward K^+ current (phase 3; Fig. 19a). The plateau and associated Ca^{2+} entry are essential for contraction; the blockade of L-type channels (e.g. **dihydropyridines**) reduces force. As the AP lasts almost as long as contraction (Fig. 19b), its *refractory period* (Chapter 5) prevents another AP being initiated until the muscle relaxes; thus, cardiac muscle cannot exhibit tetanus (Chapter 14).

The sinoatrial node and origin of the heart beat

The sinoatrial node (SA node) AP differs from that in ventricular muscle (Fig. 19c). The resting potential starts at a more positive value (approximately −60 mV) and decays steadily with time until it reaches a threshold of around −40 mV, when an AP is initiated. The upstroke of the AP is **slow**, as it is not due to the activation of fast Na^+ channels, but instead slow **L-type Ca^{2+} channels**; the SA node contains no functional fast Na^+ channels. The slow upstroke means that conduction between SA nodal myocytes is slow; this is particularly important in the **atrioventricular node** (**AV node**), which has a similar AP. The rate of decay of the SA node resting potential determines the time it takes to reach threshold and to generate another AP, and hence determines the heart rate; it is therefore called the **pacemaker potential**. The pacemaker potential decays because of a slowly reducing outward K^+ current set against inward currents. Factors that affect these currents alter the rate of decay and the time to reach threshold, and hence the heart rate, and are called **chronotropic agents**. The sympathetic transmitter, noradrenaline (norepinephrine), is *a positive chronotrope* that increases the rate of decay and thus the heart rate, whereas the parasympathetic transmitter, **acetylcholine**, lengthens the time to reach threshold and decreases the heart rate (Fig. 19d).

Action potentials elsewhere in the heart (Fig. 19e). Atria have a similar but more triangular AP compared to the ventricles. **Purkinje** fibres in the conduction system are also similar to ventricular myocytes, but have a spike (phase 1) at the peak of the upstroke, reflecting a larger Na^+ current that contributes to their fast conduction velocity (Chapter 6). Other atrial cells, the AV node, bundle of His and Purkinje system may also exhibit decaying resting potentials that can act as pacemakers. However, the SA node is normally fastest and predominates. This is called **dominance** or **overdrive suppression**.

Excitation–contraction coupling (Fig. 19f)

Contraction. Cardiac muscle contracts when intracellular Ca^{2+} rises above 100 nM. Although Ca^{2+} entry during the AP is essential for contraction, it only accounts for ~25% of the rise in intracellular Ca^{2+}. The rest is released from Ca^{2+} stores in the **sarcoplasmic reticulum** (**SR**). APs travel down invaginations of the sarcolemma called **T-tubules**, which are close to, but do not touch, the **terminal cisternae** of the SR ❶. During the AP plateau, Ca^{2+} enters the cell and activates Ca^{2+}-sensitive **Ca^{2+} release channels** in the SR ❷, allowing stored Ca^{2+} to flood into the cytosol; this is **Ca^{2+}-induced Ca^{2+} release** (CICR). The amount of Ca^{2+} released depends on how much is stored and how much Ca^{2+} enters during the AP. Modulation of the latter is a key way in which cardiac muscle force is regulated (see below). Peak intracellular $[Ca^{2+}]$ normally rises to ~2 μM, although maximum contraction occurs above 10 μM.

Relaxation. Ca^{2+} is rapidly pumped back into the SR (**sequestered**) by adenosine triphosphate (ATP)-dependent Ca^{2+} pumps (Ca^{2+} ATPase) ❸. However, Ca^{2+} that entered the myocyte during the AP must also be removed again. This is primarily performed by the **Na^+–Ca^{2+} exchanger** in the membrane, which pumps one Ca^{2+} ion out in exchange for three Na^+ ions, using the Na^+ electrochemical gradient as an energy source ❹. This is relatively slow, and continues during diastole. If the latter is shortened, i.e. when the heart rate rises, more Ca^{2+} is left inside the cell and the cardiac force increases. This is the **staircase** or Treppe effect.

Regulation of contractility: inotropic agents (Fig. 19f)

Sympathetic stimulation increases cardiac muscle **contractility** (Chapter 20) because it causes the release of noradrenaline, a **positive inotrope**. Noradrenaline binds to β_1-adrenoceptors on the membrane and causes increased Ca^{2+} entry via L-type Ca^{2+} channels during the AP ❺, and thus increases Ca^{2+} release from the SR (❷; see above). Noradrenaline also accelerates Ca^{2+} sequestration into the SR ❻. The contractility is also increased by slowing the removal of Ca^{2+} from the myocyte. **Cardiac glycosides** (e.g. *digoxin*) inhibit the Na^+ pump which removes Na^+ from the cell (Chapter 4) ❻. Intracellular $[Na^+]$ therefore increases and the Na^+ gradient across the membrane is reduced. This depresses Na^+–Ca^{2+} exchange ❹, which relies on the Na^+ gradient for its motive force, and Ca^{2+} is pumped out of the cell less rapidly. Consequently, more Ca^{2+} is available inside the myocyte for the next beat, and force increases. **Acidosis** (blood pH < 7.3) is **negatively inotropic**, largely because H^+ competes for Ca^{2+}-binding sites.

20 Control of cardiac output and Starling's law of the heart

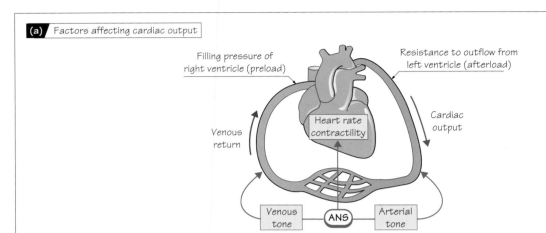

(a) Factors affecting cardiac output

Filling pressure of right ventricle (preload)

Resistance to outflow from left ventricle (afterload)

Heart rate contractility

Venous return

Cardiac output

Venous tone — ANS — Arterial tone

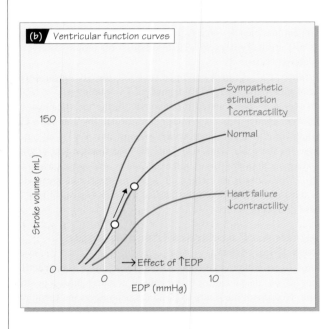

(b) Ventricular function curves

Sympathetic stimulation ↑contractility

Normal

Heart failure ↓contractility

→ Effect of ↑EDP

Stroke volume (mL) — 150, 0

EDP (mmHg) — 0, 10

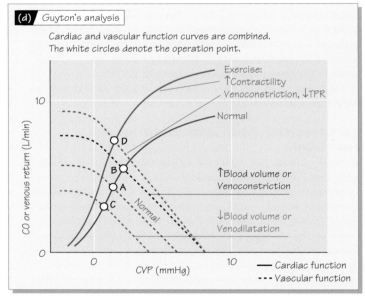

(d) Guyton's analysis

Cardiac and vascular function curves are combined.
The white circles denote the operation point.

Exercise:
↑Contractility
Venoconstriction, ↓TPR

Normal

D
B
A
C

↑Blood volume or Venoconstriction

↓Blood volume or Venodilatation

Normal

CO or venous return (L/min) — 10, 0

CVP (mmHg) — 0, 10

—— Cardiac function
- - - Vascular function

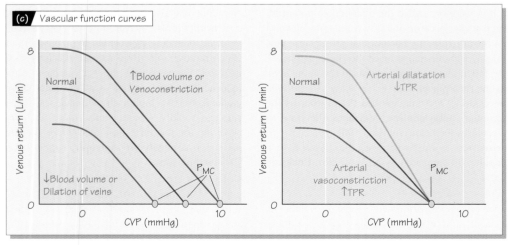

(c) Vascular function curves

Normal

↑Blood volume or Venoconstriction

↓Blood volume or Dilation of veins

P_{MC}

Venous return (L/min) — 8, 0

CVP (mmHg) — 0, 10

Normal

Arterial dilatation ↓TPR

Arterial vasoconstriction ↑TPR

P_{MC}

Venous return (L/min) — 8, 0

CVP (mmHg) — 0, 10

Physiology at a Glance, Third Edition. Jeremy P.T. Ward and Roger W.A. Linden.

© 2013 John Wiley & Sons, Ltd. Published 2013 by John Wiley & Sons, Ltd.

Cardiac output (**CO**) is determined by the heart rate and stroke volume (SV): CO = heart rate × SV. SV is influenced by the filling pressure (**preload**), cardiac muscle force, and the pressure against which the heart has to pump (**afterload**). Both the heart rate and force are modulated by the autonomic nervous system (ANS) (Fig. 20a). The heart and vasculature form a closed system, so except for transient perturbations **venous return** *must* equal CO.

Filling pressure and Starling's law

The right ventricular end-diastolic pressure (**EDP**) is dependent on central venous pressure (**CVP**); left ventricular EDP is dependent on pulmonary venous pressure. EDP and the compliance of the ventricle (how easy it is to inflate) determine the end-diastolic volume (**EDV**). As EDP (and so EDV) increases, the force of systolic contraction and thus SV also increases. This is called the **Frank–Starling** relationship, and the graph relating SV to EDP is called a **ventricular function curve** (Fig. 20b). The force of contraction is related to the degree of stretch of cardiac muscle, and **Starling's law of the heart** states: '*The energy released during contraction depends on the initial fibre length*'. As muscle is stretched, more myosin cross-bridges can form, increasing force (**sliding filament theory**; Chapter 12). However, cardiac muscle has a much steeper relationship between stretch and force than skeletal muscle, because stretch also increases the Ca^{2+} sensitivity of troponin (Chapter 12), so more force is generated for the same intracellular Ca^{2+}. The ventricular function curve is therefore steep, and small changes in EDP lead to large increases in SV.

Importance of Starling's law

The most important consequence of Starling's law is that **SV in the left and right ventricles is matched**. If, for example, right ventricular SV increases, the amount of blood in the lungs and thus pulmonary vascular pressure will also increase. As the latter determines left ventricular EDP, left ventricular SV increases due to Starling's law until it again matches that of the right ventricle, when input to and output from the lungs equalize and the pressure stops rising. This represents a rightward shift along the function curve (Fig. 20b). Starling's law thus explains how an increase in CVP, which is only perceived by the right ventricle, can increase CO. It also explains why an increase in afterload (e.g. hypertension) may have little effect on CO. It should be intuitive that an increase in afterload will reduce SV if cardiac force is not increased. However, this means more blood is left in the left ventricle after systole, and also that the outputs of the two ventricles no longer match. As a result, blood accumulates on the venous side and filling pressure rises. Cardiac force therefore increases according to Starling's law until it overcomes the increased afterload and, after a few beats, CO is restored at the expense of an increased EDP.

Autonomic nervous system

The autonomic nervous system (**ANS**) provides an important extrinsic influence on CO. **Sympathetic** stimulation increases heart rate whereas **parasympathetic** decreases it; sympathetic stimulation also increases cardiac muscle force without a change in stretch (or EDV) (i.e. it increases **contractility**; Chapter 19). The ventricular function curve therefore shifts upwards (Fig. 20b). By definition, Starling's law does *not* increase contractility.

Activation of sympathetic nerves also induces arterial and venous vasoconstriction (Chapter 22). An often overlooked point is that these differ in effect. Arterial vasoconstriction increases total peripheral resistance (**TPR**) and impedes blood flow. However, unlike arteries, veins are highly compliant (stretch easily), and contain ~70% of blood volume. Venoconstriction reduces the compliance of veins and hence their capacity (amount of blood they contain), and therefore has the same effect as increasing blood volume, i.e. **CVP increases**. Venoconstriction does not significantly impede flow because venous resistance is very low compared to TPR. Sympathetic stimulation therefore increases CO by increasing heart rate, contractility and CVP.

Postural hypotension. On standing from a prone position, gravity causes blood to pool in the legs and CVP falls. This in turn causes a fall in CO (*due to Starling's law*) and thus a fall in blood pressure. This **postural hypotension** is normally rapidly corrected by the **baroreceptor reflex** (Chapter 22), which causes venoconstriction (partially restoring CVP) and an increase in heart rate and contractility, so restoring CO and blood pressure. Even in healthy people it occasionally causes a temporary blackout (fainting or *syncope*) due to reduced cerebral perfusion. Reduction of ANS function with age accounts for a greater likelihood of postural hypotension as we get older.

Venous return and vascular function curves

Blood flow is driven by the arterial–venous pressure difference, so **venous return** will be *impeded* by a rise in CVP (Fig. 20c). This is at first glance inconsistent with Starling's law if CO *must* equal venous return. However, CVP is only altered by changes in blood volume or its distribution (e.g. venoconstriction), and these also alter the relationship between CVP and venous return (the **vascular function curve**; Fig. 20c). This figure indicates that venous return is maximum when CVP is zero (the flattening of the curve reflects venous collapse at negative pressures). Conversely, venous return will be zero if the heart stops, when pressures equalize throughout the vascular system to a **mean circulatory pressure** (P_{MC}); by definition CVP will equal P_{MC} at this point. P_{MC} is dependent on the vascular volume and compliance, and thus primarily on venous status (see above). Raising blood volume or venoconstriction therefore increases P_{MC} and causes a parallel shift of the vascular function curve; the reverse occurs in blood loss. In contrast, arterial vasoconstriction has insignificant effects on P_{MC} because the volume of resistance arteries is small; it does however reduce venous return due to the increase in TPR (see above). The net effect is therefore to reduce the slope of the curve, whilst a reduction in TPR increases it.

Guyton's analysis combines vascular and cardiac function curves into one graph (Fig. 20d). The only point where CO and venous return are equal is the intersection of the curves (A); this is thus the **operating point**. If blood volume is now increased, the shift in the vascular function curve leads to a new operating point (B) where both CO and CVP are increased; blood loss does the opposite (C). In exercise, a more complex example, sympathetic stimulation causes both increased cardiac contractility and venoconstriction, but TPR *falls* due to vasodilation in active muscle. Thus both cardiac and vascular function curves shift up, but because of the fall in TPR the latter has a steeper slope (see above). The new operating point (D) shows that in exercise CO can be greatly increased with only minor changes in CVP.

21 Blood vessels

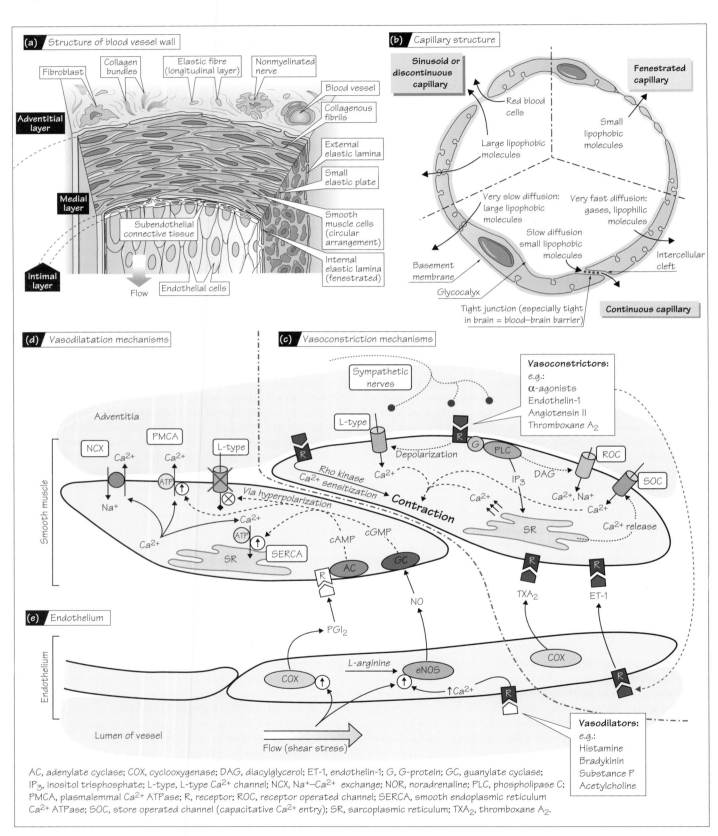

(a) Structure of blood vessel wall

(b) Capillary structure

(d) Vasodilatation mechanisms

(c) Vasoconstriction mechanisms

(e) Endothelium

AC, adenylate cyclase; COX, cyclooxygenase; DAG, diacylglycerol; ET-1, endothelin-1; G, G-protein; GC, guanylate cyclase; IP_3, inositol trisphosphate; L-type, L-type Ca^{2+} channel; NCX, Na^+–Ca^{2+} exchange; NOR, noradrenaline; PLC, phospholipase C; PMCA, plasmalemmal Ca^{2+} ATPase; R, receptor; ROC, receptor operated channel; SERCA, smooth endoplasmic reticulum Ca^{2+} ATPase; SOC, store operated channel (capacitative Ca^{2+} entry); SR, sarcoplasmic reticulum; TXA_2, thromboxane A_2.

Physiology at a Glance, Third Edition. Jeremy P.T. Ward and Roger W.A. Linden.

 © 2013 John Wiley & Sons, Ltd. Published 2013 by John Wiley & Sons, Ltd.

Structure

The walls of larger blood vessels comprise three layers: an inner **intima** (*tunica intima*) consisting of a thin layer of **endothelial cells**; a thick **media** (*tunica media*) containing **smooth muscle** and **elastin filaments** that provide elastic properties; and an outer **adventitia** (*tunica adventitia*) consisting of fibroblasts and nerves embedded in collagenous tissue (Fig. 21a). The layers are separated by inner and outer **elastic lamina**. In large vessels, the adventitia contains a network of blood vessels called the **vasa vasorum** (*vessel of vessels*) supplying the smooth muscle. Veins have a thinner media than arteries, and contain less smooth muscle. All three layers contain fibrous **collagen**, which acts as a framework to which cells are anchored.

Vascular smooth muscle cells are elongated, 15–100 μm in length, and tend to be orientated in a spiral around the vessel; the lumen therefore narrows as they contract. Cells are connected by **gap junctions**, allowing electrical coupling and depolarization to spread from cell to cell. The structure and function of smooth muscle are described in Chapter 15.

Capillaries and the smallest venules are formed from a single layer of endothelial cells supported on the outside by a 50–100-nm thick **basal lamina** containing collagen. The luminal surface is covered by a glycoprotein network called the **glycocalyx**. There are three basic types of capillary, varying in permeability (Fig. 21b). **Continuous capillaries** have a low permeability, as junctions between the endothelial cells are very tight and prevent the diffusion of lipophobic molecules of >10 000 Da. They are found in skin, lungs, central nervous system and muscle. **Fenestrated capillaries** have less tight junctions and the endothelial cells are also punctured by 50–100-nm pores (**fenestrae**); they are therefore much more permeable. They are found where large amounts of fluid or material need to diffuse across the capillary wall, including endocrine glands, renal glomeruli and intestinal villa. **Discontinuous capillaries** are found in bone marrow, liver and spleen, and have gaps large enough for red blood cells to pass through. The microcirculation is discussed further in Chapter 23.

Regulation of function and excitation–contraction coupling

Vasoconstriction (Fig. 21c). Most vasoconstrictors bind to receptors and cause a guanosine triphosphate-binding protein (G-protein)-mediated elevation in intracellular $[Ca^{2+}]$, leading to contraction. Important vasoconstrictors include endothelin-1, angiotensin II (Chapter 35) and the sympathetic transmitter noradrenaline (norepinephrine) (Chapter 7).

Ca^{2+} *release*. Binding to a receptor activates **phospholipase C**, which generates the second messengers inositol trisphosphate (**IP$_3$**) and diacylglycerol (**DAG**) from membrane phospholipids. IP$_3$ binds to receptors on the **sarcoplasmic reticulum** (SR) causing Ca^{2+} channels to open and Ca^{2+} to flood into the cytoplasm. This response may only be transient as the store rapidly empties, but may initiate *capacitative Ca^{2+} entry* (see below).

Ca^{2+} *entry*. Vasoconstrictors also cause depolarization, which activates Ca^{2+} entry via **L-type voltage-gated Ca^{2+} channels** as in cardiac muscle (Chapter 19). Unlike cardiac muscle, most types of vascular smooth muscle do not generate action potentials, but instead depolarization is graded, allowing graded entry of Ca^{2+}. **Receptor operated channels** (**ROC**) may also be activated, some by DAG, through which both Ca^{2+} and Na^+ can enter the cell; the latter contributes to depolarization. IP$_3$-stimulated emptying of Ca^{2+} stores can also directly activate **store operated channels** (**SOC**) in the membrane, causing **capacitative Ca^{2+} entry**.

Importantly, many agonists also cause **Ca^{2+} sensitization** of the contractile apparatus, i.e. more force for the same rise in Ca^{2+}. This is mediated by **Rho kinase**, although protein kinase C, which is activated by DAG, may also be involved. The relative importance of the above mechanisms depends on the vascular bed and vasoconstrictor. In small-resistance arteries, depolarization and voltage-gated Ca^{2+} entry are probably most important. Most systemic arteries exhibit a degree of **basal** (*myogenic*) **tone** in the absence of vasoconstrictors.

Ca^{2+} *removal and vasodilatation* (Fig. 21d). Ca^{2+} is pumped back into the SR (*sequestrated*) by the **smooth endoplasmic reticulum Ca^{2+} ATPase** (**SERCA**) which can rapidly reduce cytosolic Ca^{2+}. Ca^{2+} is also removed from the cell by a **plasma membrane Ca^{2+} ATPase** (PMCA) and **Na^+–Ca^{2+} exchange** (NCX; Chapter 19). Most endogenous vasodilators cause relaxation by increasing cyclic guanosine monophosphate (cGMP) (e.g. **nitric oxide**, NO) or cyclic adenosine monophosphate (cAMP) (e.g. **prostacyclin**, β-adrenergic receptor agonists). These second messengers act via protein kinase G (PKG) or protein kinase A (PKA), respectively. Both PKG and PKA lower intracellular Ca^{2+}, partly by stimulating SERCA and PMCA, and partly by hyperpolarizing the membrane (i.e. so voltage-gated Ca^{2+} entry is inhibited). L-type Ca^{2+} channel blocker drugs, such as **verapamil** or **dihydropyridines**, are clinically effective vasodilators.

The endothelium (Fig. 21e)

The endothelium plays a crucial role in the regulation of vascular tone. In response to substances in the blood or changes in blood flow, it can synthesize several important vasodilators, including **NO** (*endothelium-derived relaxing factor*, EDRF) and **prostacyclin** (prostaglandin I$_2$, PGI$_2$), as well as potent vasoconstrictors, such as **endothelin-1** and **thromboxane A$_2$** (TXA$_2$).

NO is synthesized by the endothelial **nitric oxide synthase** (eNOS) from L-arginine. eNOS activity and NO production are increased by factors that elevate intracellular Ca^{2+}, including local mediators such as **bradykinin**, **histamine** and **serotonin**, and some neurotransmitters (e.g. **substance P**). Increased flow (**shear stress**) also stimulates NO production, and additionally activates prostacyclin synthesis. The basal production of NO continuously modulates vascular resistance, as it has been found that inhibition of eNOS causes the blood pressure to rise. NO also inhibits platelet activation and **thrombosis** (inappropriate clotting) (Chapter 9).

Endothelin-1 is an extremely potent vasoconstrictor peptide which is released from the endothelium in the presence of many other vasoconstrictors, including angiotensin II, antidiuretic hormone (ADH; *vasopressin*) and noradrenaline, and may be increased in disease and hypoxia. As endothelin receptor blockade causes a fall in the peripheral resistance of healthy humans, it seems to contribute to the maintenance of blood pressure.

The **eicosanoids** prostacyclin and TXA$_2$ are synthesized by the cyclooxygenase pathway from arachidonic acid, which is made from membrane phospholipids by phospholipase A$_2$. In most vessels prostacyclin is most important.

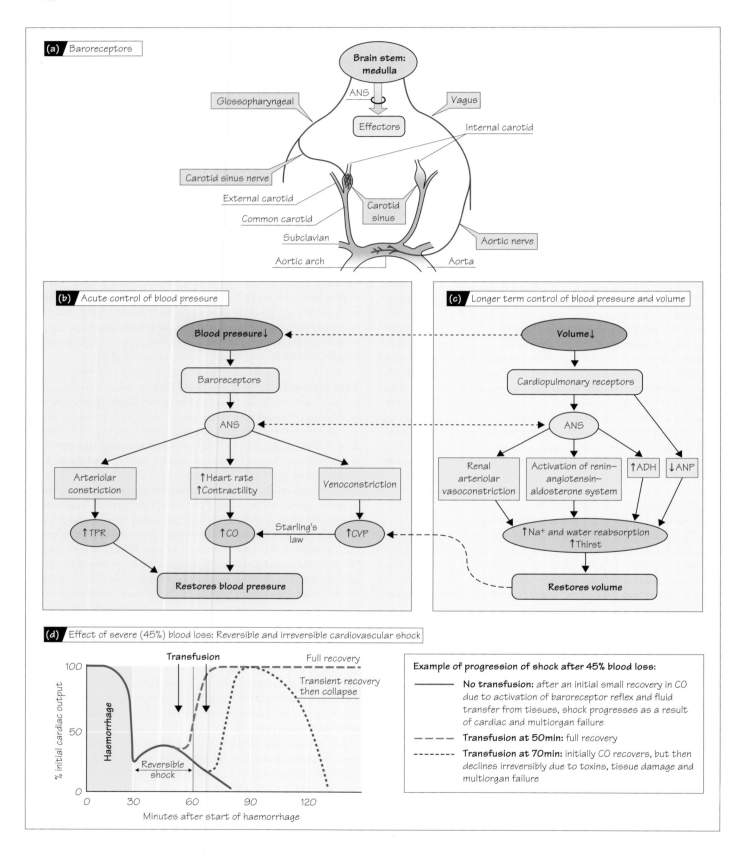

(a) Baroreceptors

(b) Acute control of blood pressure

(c) Longer term control of blood pressure and volume

(d) Effect of severe (45%) blood loss: Reversible and irreversible cardiovascular shock

Example of progression of shock after 45% blood loss:

— **No transfusion:** after an initial small recovery in CO due to activation of baroreceptor reflex and fluid transfer from tissues, shock progresses as a result of cardiac and multiorgan failure

– – **Transfusion at 50min:** full recovery

· · · · **Transfusion at 70min:** initially CO recovers, but then declines irreversibly due to toxins, tissue damage and multiorgan failure

Physiology at a Glance, Third Edition. Jeremy P.T. Ward and Roger W.A. Linden.

© 2013 John Wiley & Sons, Ltd. Published 2013 by John Wiley & Sons, Ltd.

Tissues can independently alter their blood flow by changing their vascular resistance. So that this does not have a knock-on effect elsewhere, the pressure head provided by the mean arterial blood pressure (MAP) must be controlled. MAP is determined by the **total peripheral resistance** (TPR) and **cardiac output** (MAP = cardiac output × TPR), which is itself dependent on the **central venous pressure** (CVP) (Chapter 20). CVP is highly dependent on the **blood volume**. Alterations of any of these variables may change MAP.

Effect of gravity. When standing, the blood pressure at the ankle is ~90 mmHg higher than that at the level of the heart, due to the weight of the column of blood between the two. Similarly, the pressure in the head is ~30 mmHg less than that at the level of the heart. Blood pressure is always measured at the level of the heart. Gravity does not affect the driving force between arteries and veins because arterial and venous pressures are affected equally.

Acute regulation of the mean arterial blood pressure: the baroreceptor reflex

Physiological regulation commonly involves **negative feedback**. This requires a **sensor** that detects the controlled variable (e.g. MAP), a **comparator** that compares the sensor output to a **set point**, and a **feedback pathway** driving **effectors** that adjust the variable until the difference between the sensor output and the set point is minimized (Chapter 1). The sensor for MAP is provided by **baroreceptors** (stretch receptors) located in the **carotid sinus** and **aortic arch** (Fig. 22a). A decrease in MAP reduces arterial wall stretch and *decreases* baroreceptor activity, resulting in decreased firing in afferent nerves travelling via the glossopharyngeal and vagus to the medulla of the brain stem, where the activity of the **autonomic nervous system** (**ANS**) (Chapter 7) is coordinated. Sympathetic nervous activity consequently *increases*, causing an increased heart rate and cardiac contractility (Chapter 20), peripheral vasoconstriction, and an increase in TPR and venoconstriction, which increases CVP (Chapter 21). Parasympathetic activity *decreases*, contributing to the rise in heart rate (Chapter 19). MAP therefore returns to normal (Fig. 22b). An increase in MAP has the opposite effects.

The baroreceptors are most sensitive between 80 and 150 mmHg, and their sensitivity is increased by a large **pulse pressure** (Chapter 16). They also show **adaptation**; if a new pressure is maintained for a few hours, activity slowly returns towards (but not to) normal. The baroreceptor reflex is important for buffering *short-term* changes in MAP, e.g. when muscle blood flow increases rapidly in exercise. Cutting the baroreceptor nerves has a minor effect on average MAP, but fluctuations in pressure are much greater.

Posture. Changes in posture provide a good example of the acute baroreceptor reflex. When standing from a supine position, blood pools in the veins of the legs, causing a fall in CVP; cardiac output and MAP therefore fall (**postural hypotension**; Chapter 20). Baroreceptor firing is reduced and the baroreceptor reflex is activated. *Venoconstriction* reduces blood pooling and helps restore CVP which, coupled with an *increase in heart rate and cardiac contractility*, returns cardiac output towards normal; peripheral *vasoconstriction* assists the restoration of MAP. The transient dizziness or blackout (**syncope**) occasionally experienced when rising rapidly is due to a fall in cerebral perfusion that occurs before cardiac output and MAP can be corrected.

Long-term regulation: control of blood volume (Fig. 22c)

The blood volume is dependent on total body Na^+ and water. These are regulated by the kidneys, and it is therefore strongly recommended that this chapter is read together with Chapter 35, where the renal mechanisms involved are discussed in detail.

The activation of the baroreceptor reflex by a reduction in MAP leads to renal arteriolar constriction mediated by efferent sympathetic nerves. This and the fall in MAP itself cause a reduction in renal perfusion pressure, which reduces glomerular filtration and so inhibits excretion of Na^+ and water in the urine. Sympathetic stimulation and reduced arteriolar pressure also activate the **renin–angiotensin system** (Chapter 35) and thus the production of **angiotensin II**, a potent vasoconstrictor that increases TPR. Angiotensin II also stimulates the production of **aldosterone** from the adrenal cortex, which promotes renal Na^+ reabsorption. The net effect is Na^+ and water retention, and an increase in blood volume (Fig. 22d). Conversely, a rise in MAP increases Na^+ and water excretion.

Changes in blood volume are sensed directly by **cardiopulmonary receptors: veno-atrial receptors** are located around the join between the veins and atria, and **atrial receptors** in the atrial wall. These effectively respond to changes in CVP and *blood volume*. Stimulation (stretch) suppresses the renin–angiotensin system, sympathetic activity and secretion of **antidiuretic hormone** (ADH, vasopressin), but increases release of **atrial natriuretic peptide** (ANP) from the atria. Together, these changes promote renal Na^+ and water excretion and reduce blood volume (Chapters 34 and 35). A fall in blood volume will induce the opposite effects. The cardiopulmonary receptors normally cause **tonic depression** – cutting their efferent nerves increases the heart rate and causes vasoconstriction in the gut, kidney and skeletal muscle, thus raising MAP.

Cardiovascular shock and haemorrhage

Cardiovascular shock. This is an acute condition with inadequate blood flow throughout the body, commonly associated with a fall in MAP. It can result from reduced blood volume (**hypovolumic shock**), profound vasodilatation (**low-resistance shock**) or acute failure of the heart to pump (**cardiogenic shock**). The most common cause of hypovolumic shock is **haemorrhage**; others include severe burns, vomiting and diarrhoea (e.g. cholera). Low-resistance shock is due to the profound vasodilatation caused by bacterial infection (**septic shock**) or powerful allergic reactions (e.g. to bee stings or peanuts; **anaphylactic shock**).

Haemorrhage. Some 20% of the blood volume can be lost without significant problems, as the baroreceptor reflex mobilizes blood from capacitance vessels and maintains MAP. Volume is restored within 24 h because arteriolar constriction reduces the capillary pressure and fluid moves from tissues into the plasma (Chapter 23), urine production is suppressed (see above) and ADH and angiotensin II stimulate thirst. Greater loss (30–50%) can be survived, but only with transfusion within ~1 h (the '**golden hour**') (Fig. 22d). After this, **irreversible shock** generally develops, which is irretrievable even with transfusion. This is because the reduced MAP and consequent profound peripheral vasoconstriction cause tissue ischaemia and the build-up of toxins and acidity, which damage the microvasculature and heart and lead to *multiorgan failure*.

23 The microcirculation, filtration and lymphatics

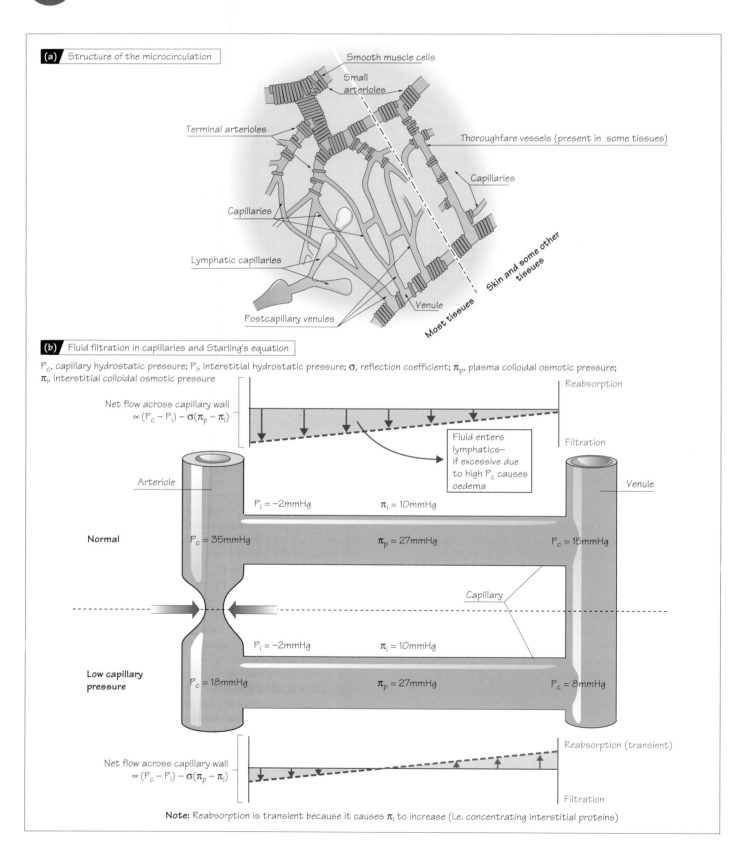

(a) Structure of the microcirculation

Smooth muscle cells
Small arterioles
Terminal arterioles
Thoroughfare vessels (present in some tissues)
Capillaries
Capillaries
Lymphatic capillaries
Postcapillary venules
Venule
Skin and some other tissues
Most tissues

(b) Fluid filtration in capillaries and Starling's equation

P_c, capillary hydrostatic pressure; P_i, interstitial hydrostatic pressure; σ, reflection coefficient; π_p, plasma colloidal osmotic pressure; π_i, interstitial colloidal osmotic pressure

Net flow across capillary wall $\propto (P_c - P_i) - \sigma(\pi_p - \pi_i)$

Reabsorption

Filtration

Fluid enters lymphatics— if excessive due to high P_c causes oedema

Arteriole

Venule

$P_i = -2$mmHg $\pi_i = 10$mmHg

Normal $P_c = 35$mmHg $\pi_p = 27$mmHg $P_c = 15$mmHg

Capillary

$P_i = -2$mmHg $\pi_i = 10$mmHg

Low capillary pressure $P_c = 18$mmHg $\pi_p = 27$mmHg $P_c = 8$mmHg

Net flow across capillary wall $\propto (P_c - P_i) - \sigma(\pi_p - \pi_i)$

Reabsorption (transient)

Filtration

Note: Reabsorption is transient because it causes π_i to increase (i.e. concentrating interstitial proteins)

Physiology at a Glance, Third Edition. Jeremy P.T. Ward and Roger W.A. Linden.
 © 2013 John Wiley & Sons, Ltd. Published 2013 by John Wiley & Sons, Ltd.

The **microcirculation** is perhaps the *raison d'être* for the cardiovascular system, as it is here that exchange between blood and tissues occurs. It consists of the smallest (**terminal**) **arterioles** and the **exchange vessels – capillaries** and **small venules** (Chapter 16). Blood flow into the microcirculation is regulated by the vasoconstriction of small arterioles, activated by sympathetic stimulation through numerous nerve endings in their walls (Chapters 7 and 22). Each small arteriole feeds many capillaries via several **terminal arterioles** (Fig. 23a), which are not innervated. Instead, the vasoconstriction of terminal arterioles is mediated by **local metabolic products** (Chapter 24), allowing perfusion to be matched to metabolism. A *few* tissues (e.g. mesenteric, skin) have **thoroughfare vessels** connecting small arterioles and venules directly. Note that the term '*pre-capillary sphincter*' is misleading and should be avoided, as no such anatomical structures exist.

Transcapillary exchange

Water, gases and other substances cross the capillary wall mainly by **diffusion** down their concentration gradients (Chapter 11). O_2 and CO_2 are highly **lipophilic** (soluble in lipids), and can cross the endothelial lipid bilayer membrane easily. This is, however, impermeable to **hydrophilic** ('water-loving', lipid-insoluble) molecules, such as glucose, and **polar** (charged) molecules and ions (electrolytes). Such substances mainly cross the wall of **continuous capillaries** through the gaps between endothelial cells. This is slowed by **tight junctions** between cells and by the **glycocalyx** (Chapter 21), so that diffusion is 1000–10000 times slower than for lipophilic substances. This **small pore** system also prevents the diffusion of substances greater than 10000 Da (e.g. plasma proteins). The latter can cross the capillary wall, but extremely slowly; this may involve **large pores** through endothelial cells. **Fenestrated capillaries** (gut, joints, kidneys) are 10-fold more permeable than continuous capillaries because of pores called **fenestrae** (from the Latin for '*windows*'), whereas **discontinuous capillaries** are highly permeable due to large spaces between endothelial cells, and occur where red cells need to cross the capillary wall (bone marrow, spleen, liver) (Chapter 21).

Filtration (Fig. 23b)

The capillary walls are much more permeable to water and electrolytes than to proteins (see above). The concentration of electrolytes (e.g. Na^+, Cl^-), and therefore the osmotic pressure exerted by them (**crystalloid osmotic pressure**), is very similar in plasma and interstitial fluid, and has little effect on fluid movement. The protein concentration in plasma however is greater than that in interstitial fluid, and the component of osmotic pressure exerted by proteins (**colloidal osmotic** or **oncotic pressure**) in the plasma (~27 mmHg) is therefore greater than in the interstitial fluid (~10 mmHg). Water tends to flow from a *low* to a *high* osmotic pressure, but from a *high* to a *low* hydrostatic pressure. The net flow of water across the capillary wall is therefore determined by the balance between the hydrostatic (P) and colloidal osmotic (π) pressures, according to **Starling's equation**, flow $\propto$ $(P_c - P_i) - \sigma(\pi_p - \pi_i)$, where $(P_c - P_i)$ is the difference in hydrostatic pressure between capillary and interstitial fluid, and $(\pi_p - \pi_i)$ is the difference in colloidal osmotic pressure between plasma and interstitial fluid; $(\pi_p - \pi_i)$ has an average value of ~17 mmHg. σ is the **reflection coefficient** (~0.9), a measure of how difficult it is for plasma proteins to cross the capillary wall. Note that the interstitial protein concentration, and therefore π_i, differs between tissues; in the lung for example $(\pi_p - \pi_i)$ is ~13 mmHg.

The capillary hydrostatic pressure normally varies from ~35 mmHg at the arteriolar end to ~15 mmHg at the venous end, whereas the interstitial hydrostatic pressure is approximately –2 mmHg. $(P_c - P_i)$ is therefore greater than $\sigma(\pi_p - \pi_i)$ along the length of the capillary, resulting in the **net filtration** of water into the interstitial space (Fig. 23b). Although arteriolar constriction will reduce capillary pressure and therefore lead to the reabsorption of fluid, this will normally be transient due to the concentration of interstitial fluid (i.e. increased π_i). A reduction in plasma protein (e.g. *starvation*), or a loss of endothelium integrity and thus diffusion of protein into the interstitial space (e.g. *severe inflammation*, *ischaemia*), will similarly reduce $(\pi_p - \pi_i)$, leading to enhanced filtration and loss of fluid into the tissues. This is also caused by a high venous pressure (**oedema**; see below).

Lymphatics

Fluid filtered by the microcirculation (~8 L per day) is returned to the blood by the **lymphatic system**. Lymphatic capillaries are blind-ended bulbous tubes (diameter, ~15–75 µm) walled with endothelial cells (Fig. 23a). These allow the entry of fluid, proteins and bacteria, but prevent their exit. Lymphatic capillaries merge into **collecting lymphatics** and then larger lymphatic vessels, both containing smooth muscle and **unidirectional valves**. Lymph is propelled, by smooth muscle constriction and compression of the vessels by body movement, into **afferent lymphatics** and then the **lymphatic nodes**, where bacteria and other foreign materials are removed by phagocytes. Most fluid is re-absorbed here by capillaries, with the remainder returning via **efferent lymphatics** and the thoracic duct into the subclavian veins. Lymphatics are also important for lipid absorption in the gut.

Oedema

Oedema is swelling of the tissues due to excess fluid in the interstitial space. It is caused when filtration is increased to the extent that the lymphatics are unable to remove the fluid fast enough (see above), or by dysfunctional lymphatic drainage (e.g. *elephantiasis*, the blockage of lymphatics with filarial nematode worms). **Inflammation** (Chapter 10) causes swelling and oedema because it increases capillary permeability, allowing protein to leak into the interstitium and disrupt the oncotic pressure gradient, so filtration is increased. Reduced venous drainage (increased venous pressure) also increases filtration and can lead to oedema; standing without moving the legs prevents the operation of the **muscle pump** (Chapter 16), local venous pressure rises, and the legs swell. In **congestive heart failure**, reduced cardiac function results in increased pulmonary and central venous pressure (Chapter 20), leading, respectively, to **pulmonary oedema** (alveoli fill with fluid) and **peripheral oedema** [swelling of the legs and liver, and accumulation of fluid in the peritoneum (*ascites*)]. Severe protein starvation can cause generalized oedema and a grossly swollen abdomen due to ascites and an enlarged liver (*kwashiorkor*).

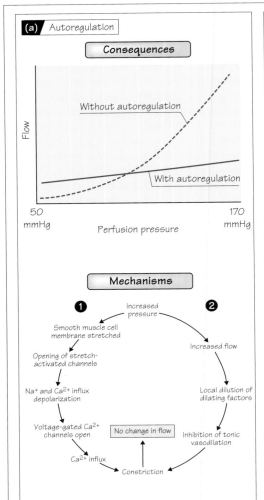

(a) Autoregulation

Consequences

Flow vs Perfusion pressure (50 mmHg to 170 mmHg):
- Without autoregulation
- With autoregulation

Mechanisms

① Increased pressure → Smooth muscle cell membrane stretched → Opening of stretch-activated channels → Na⁺ and Ca²⁺ influx depolarization → Voltage-gated Ca²⁺ channels open → Ca²⁺ influx → Constriction

② Increased pressure → Increased flow → Local dilution of dilating factors → Inhibition of tonic vasodilation → Constriction

No change in flow

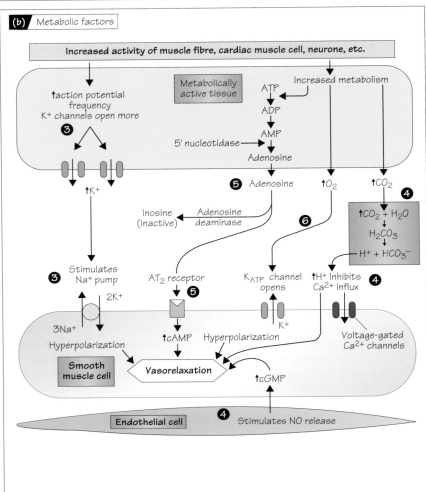

(b) Metabolic factors

Increased activity of muscle fibre, cardiac muscle cell, neurone, etc.

③ ↑action potential frequency K⁺ channels open more → ↑K⁺

Metabolically active tissue

Increased metabolism → ATP → ADP → AMP → (5' nucleotidase) → Adenosine

⑤ Adenosine → Inosine (inactive) ← Adenosine deaminase

↑O₂ ↑CO₂

④ ↑CO₂ + H₂O → H₂CO₃ → H⁺ + HCO₃⁻

③ Stimulates Na⁺ pump — 3Na⁺ / 2K⁺ — Hyperpolarization

⑤ AT₂ receptor → ↑cAMP → Hyperpolarization

K_ATP channel opens → K⁺

④ ↑H⁺ Inhibits Ca²⁺ influx → Voltage-gated Ca²⁺ channels

Smooth muscle cell → Vasorelaxation

↑cGMP

Endothelial cell ④ Stimulates NO release

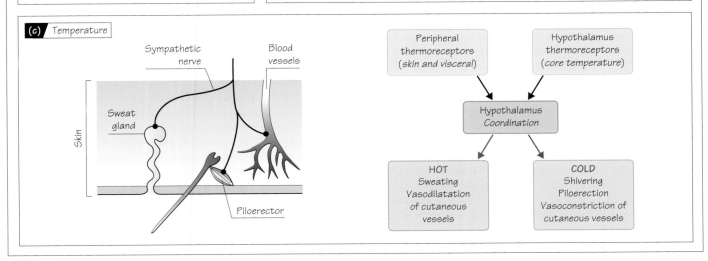

(c) Temperature

Sympathetic nerve, Blood vessels, Sweat gland, Piloerector, Skin

Peripheral thermoreceptors (skin and visceral) → Hypothalamus Coordination ← Hypothalamus thermoreceptors (core temperature)

HOT: Sweating, Vasodilatation of cutaneous vessels

COLD: Shivering, Piloerection, Vasoconstriction of cutaneous vessels

Physiology at a Glance, Third Edition. Jeremy P.T. Ward and Roger W.A. Linden.

58 © 2013 John Wiley & Sons, Ltd. Published 2013 by John Wiley & Sons, Ltd.

Local control of blood flow

In addition to the central control of blood pressure and cardiac output, tissues need to be able to regulate their own blood flow to match their requirements. This is provided by **autoregulation**, **metabolic factors** and **autocoids** (local hormones).

Autoregulation (Fig. 24a). Autoregulation is the ability to maintain a constant flow in the face of variations in pressure between ~50 and 170 mmHg. It is particularly important in the brain, kidney and heart. Two mechanisms contribute to autoregulation. The **myogenic response** ❶ involves arteriolar constriction in response to stretching of the vessel wall, probably due to activation of smooth muscle **stretch-activated Ca^{2+} channels** and Ca^{2+} entry. A reduction in pressure and stretch closes these channels, causing vasorelaxation. The second mechanism is due to locally produced **vasodilating factors** ❷. An increase in blood flow dilutes these factors, causing vasoconstriction, whereas decreased blood flow allows accumulation, causing vasodilatation.

Metabolic factors (Fig. 24b). Many factors may contribute to **metabolic hyperaemia** (*increased blood flow*). The most important are K^+, CO_2 and **adenosine**, and, in some tissues, **hypoxia**. K^+ ❸ is released from active tissues and in ischaemia; local concentrations can increase to >10 mM. It causes relaxation, partly by stimulating the **Na^+ pump**, thus both increasing Ca^{2+} removal by the Na^+–Ca^{2+} exchanger and hyperpolarizing the cell (Chapter 21). The vasodilatory effects of increased CO_2 (**hypercapnia**) and **acidosis** ❹ are mediated largely through increased **nitric oxide** production (Chapter 21) and inhibition of smooth muscle Ca^{2+} entry. **Adenosine** ❺ is a potent vasodilator released from heart, skeletal muscle and brain during increased metabolism and hypoxia. It is produced from adenosine monophosphate (AMP), a breakdown product of adenosine triphosphate (ATP), and acts by stimulating the production of cyclic AMP (**cAMP**) in smooth muscle (Chapter 21). **Hypoxia** may reduce ATP sufficiently for K_{ATP} channels to activate ❻, causing hyperpolarization.

Autocoids are mostly important in special circumstances; two examples are given. In **inflammation**, mediators such as **histamine** and **bradykinin** cause vasodilatation and increase the permeability of exchange vessels, leading to swelling, but allowing access by leucocytes and antibodies to damaged tissues. The **activation of platelets** during clotting releases the vasoconstrictors **serotonin** and **thromboxane A_2**, so reducing blood loss (Chapter 9).

Special circulations

Skeletal muscle. This comprises ~50% of the body weight and, at rest, takes 15–20% of cardiac output; during exercise, this can rise to >80%. Skeletal muscle provides a major contribution to the **total peripheral resistance**, and sympathetic regulation of muscle blood flow is important in the **baroreceptor reflex**. At rest, most capillaries are not perfused, as their arterioles are constricted. Capillaries are **recruited** during exercise by **metabolic hyperaemia**, caused by the release of K^+ and CO_2 from the muscle, and **adenosine**. This *overrides* sympathetic vasoconstriction in working muscle; the latter reduces flow in non-working muscle, *conserving* cardiac output. It should be noted that muscular contraction compresses blood vessels and inhibits flow; in rhythmic (**phasic**) activity, metabolic hyperaemia compensates by vastly increasing flow during the relaxation phase. In **isometric** (static) contractions, reduced flow can cause **muscle fatigue**.

Brain. The occlusion of blood flow to the brain causes unconsciousness within minutes. The brain receives ~15% of cardiac output, and has a high capillary density. The endothelial cells of these capillaries have very tight junctions, and contain membrane transporters that control the movement of substances, such as ions, glucose and amino acids, and tightly regulate the composition of the cerebrospinal fluid. This arrangement is called the **blood–brain barrier**, and is continuous except where substances need to be absorbed or released from the blood (e.g. pituitary gland, choroid plexus). It can cause problems for the delivery of drugs to the brain, particularly antibiotics. The **autoregulation** of cerebral blood flow is highly developed, maintaining a constant flow for blood pressures between 50 and 170 mmHg. CO_2 and K^+ are particularly important metabolic regulators in the brain, increases causing a **functional hyperaemia** linking blood flow to activity. Hyperventilation reduces blood P_{CO_2}, and can cause fainting due to cerebral vasoconstriction.

Coronary circulation. The heart has a high metabolic demand and a dense capillary network. It can extract an unusually high proportion of oxygen from the blood (~70%). In exercise, the reduced diastolic interval (Chapter 18) and increased oxygen consumption demand a greatly increased blood flow, which is achieved under the influence of **adenosine**, K^+ and **hypoxia**. The heart therefore controls its own blood flow by a well-developed **metabolic hyperaemia**. This *overrides* vasoconstriction mediated by sympathetic nerves (Chapters 7 and 21), and is assisted by circulating adrenaline (epinephrine) which causes vasodilatation via β_2-adrenergic receptors.

Skin (Fig. 24c). The main function of the cutaneous circulation is **thermoregulation**. Thoroughfare vessels (Chapter 23), formed from coiled **arteriovenous anastomoses** (AVAs), directly link arterioles and venules, allowing a high blood flow into the **venous plexus** and the radiation of heat. AVAs are found mostly in the hands, feet and areas of the face. Temperature is sensed by peripheral thermoreceptors and in the **hypothalamus**, which coordinates the response. When temperature is low, sympathetic stimulation causes the vasoconstriction of cutaneous vessels; this also occurs following activation of the *baroreceptor reflex* by low blood pressure (e.g. *pale skin* in haemorrhage and shock) (Chapter 22). **Piloerection** (raising of skin hair, 'goosebumps') traps insulating air. Increased temperatures reduce sympathetic *adrenergic* stimulation, causing vasodilatation, whereas activation of **sympathetic *cholinergic* fibres** promotes **sweating** and release of **bradykinin**, which also causes vasodilatation. The net increase in blood flow may be 30-fold.

Pulmonary circulation. The pulmonary circulation is not controlled by either autonomic nerves or metabolic products, and the most important mechanism regulating flow is **hypoxic pulmonary vasoconstriction**, in which small arteries *constrict* to hypoxia. This is unique to the lung; hypoxia causes vasodilatation elsewhere (see above). Hypoxic pulmonary vasoconstriction diverts blood away from poorly ventilated areas of the lung, thus maintaining optimal **ventilation–perfusion matching** (Chapter 30); conversely, *global hypoxia* due to lung disease or altitude detrimentally increases the pulmonary artery pressure (*pulmonary hypertension*). The pulmonary capillary pressure is normally low (~7 mmHg), but fluid filtration still occurs because the interstitial pressure is low (approximately −4 mmHg) and the interstitial colloidal osmotic pressure is high (18 mmHg) (Chapter 23).

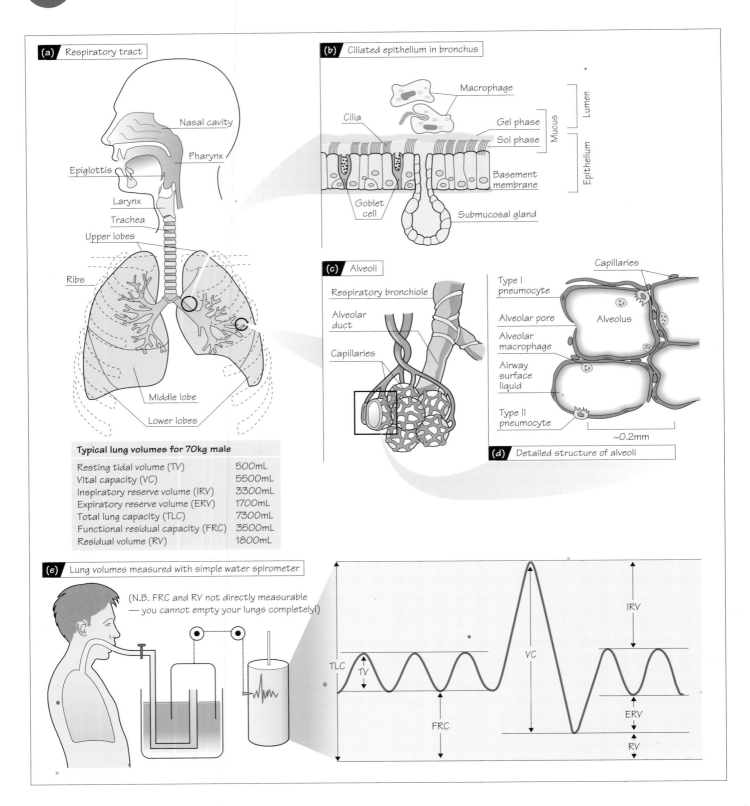

(a) Respiratory tract

Nasal cavity
Pharynx
Epiglottis
Larynx
Trachea
Upper lobes
Ribs
Middle lobe
Lower lobes

(b) Ciliated epithelium in bronchus

Macrophage
Cilia
Gel phase
Sol phase
Mucus
Lumen
Basement membrane
Epithelium
Goblet cell
Submucosal gland

(c) Alveoli

Respiratory bronchiole
Alveolar duct
Capillaries

Capillaries
Type I pneumocyte
Alveolar pore
Alveolus
Alveolar macrophage
Airway surface liquid
Type II pneumocyte
~0.2mm

(d) Detailed structure of alveoli

Typical lung volumes for 70kg male

Resting tidal volume (TV)	500mL
Vital capacity (VC)	5500mL
Inspiratory reserve volume (IRV)	3300mL
Expiratory reserve volume (ERV)	1700mL
Total lung capacity (TLC)	7300mL
Functional residual capacity (FRC)	3500mL
Residual volume (RV)	1800mL

(e) Lung volumes measured with simple water spirometer

(N.B. FRC and RV not directly measurable — you cannot empty your lungs completely!)

TLC
TV
FRC
IRV
VC
ERV
RV

Physiology at a Glance, Third Edition. Jeremy P.T. Ward and Roger W.A. Linden.

 © 2013 John Wiley & Sons, Ltd. Published 2013 by John Wiley & Sons, Ltd.

The **upper respiratory tract** includes the nose, pharynx and larynx; the **lower respiratory tract** starts at the trachea (Fig. 25a). The two **lungs** are enclosed within the **thoracic cage**, formed from the ribs, sternum and vertebral column, with the dome-shaped **diaphragm** separating the thorax from the abdomen. The left lung has two lobes, the right three. The airways, blood vessels and lymphatics enter each lung at the lung root or **hilum**, where the **pulmonary nerve plexus** receives autonomic nerves from the vagus and sympathetic trunk. The vagus contains sensory afferents from **lung receptors** (Chapter 29) and bronchoconstrictor parasympathetic efferents leading to the airways; sympathetic nerves are *bronchodilatory* (Chapter 7). Each lung lobe is made up of several wedge-shaped **bronchopulmonary segments** supplied by their own segmental bronchus, artery and vein. The lungs are covered by a thin membrane (**visceral pleura**), continuous with the **parietal pleura** that lines the inside surface of the thoracic cage. The tiny space between the pleura is filled with lubricating **pleural fluid**.

Airways (Fig. 25a)

The **trachea** divides into two **main bronchi**; their walls contain U-shaped cartilage segments linked by smooth muscle. On entering the lung, the bronchi divide repeatedly into lobar, segmental (generations 3 and 4) and small (generations 5–11) bronchi, the smallest having a diameter of ~1 mm. These all have irregular cartilaginous plates and helical bands of smooth muscle. **Bronchioles** (generations 12–16) lack cartilage and are held open by surrounding lung tissue. The smallest (**terminal**) bronchioles lead to **respiratory bronchioles** (generations 17–19), and thence to **alveolar ducts** and **sacs** (generation 23), the walls of which form **alveoli** and contain only epithelial cells (Fig. 25c,d). Small pores (**alveolar pores**, *pores of Kohn*) allow pressure equalization between alveoli. Adult human lungs contain ~17 million branches and ~300 million alveoli, providing an exchange surface of ~85 m². The **bronchial circulation** supplies airways down to the terminal bronchioles; respiratory bronchioles and below obtain nutrients from the **pulmonary circulation** (Chapter 16).

Epithelium and airway clearance

The airways from the trachea to the respiratory bronchioles are lined with **ciliated columnar epithelial cells**. **Goblet cells** and **submucosal glands** secrete a 10–15-µm thick, gel-like **mucus** that floats on a more fluid *sol phase* (Fig. 25b). Synchronous beating of the cilia moves the mucus and associated debris to the mouth (**mucociliary clearance**). Factors that increase the thickness or viscosity of the mucus (e.g. *asthma, cystic fibrosis*) or reduce cilia activity (e.g. *smoking*) impair mucociliary clearance and lead to recurrent infections. Mucus contains substances that protect the airways from pathogens (e.g. *antitrypsins, lysozyme, immunoglobulin A*).

Epithelial cells forming the walls of the alveoli and alveolar ducts are unciliated, and largely very thin **type I alveolar pneumocytes** (alveolar cells; *squamous epithelium*) (Fig. 25d). These form the gas exchange surface with the capillary endothelium (**alveolar–capillary membrane**). A few **type II pneumocytes** secrete **surfactant** which reduces the surface tension and prevents alveolar collapse (Chapter 26). **Macrophages** (*mobile phagocytes*) in the airways ingest foreign materials and destroy bacteria; in the alveoli, they take the place of cilia by clearing debris.

Respiratory muscles

The main respiratory muscles are inspiratory, the most important being the **diaphragm**; contraction pulls down the dome, reducing pressure in the thoracic cavity, and thus drawing air into the lungs. The **external intercostal muscles** assist by elevating the ribs and increasing the dimensions of the thoracic cavity. Quiet breathing is normally diaphragmatic; **accessory inspiratory muscles** (e.g. *scalene, sternomastoids*) aid inspiration if airway resistance or ventilation is high. Expiration is achieved by *passive recoil* of the lungs and chest wall, but, at high ventilation rates, this is assisted by the contraction of **abdominal muscles** which speed recoil of the diaphragm by raising abdominal pressure (e.g. exercise).

Lung volumes and pressures (Fig. 25e)

The **tidal volume** (TV) is the volume of air drawn into and out of the lungs during normal breathing; the **resting tidal volume** is normally ~500 mL but, like all lung volumes, is dependent on age, sex and height. The **vital capacity** (VC) is the maximum tidal volume, when an individual breathes in and out as far as possible. The difference in volume between a resting and maximum expiration is the **expiratory reserve volume** (ERV); the equivalent for inspiration is the **inspiratory reserve volume** (IRV). The volume in the lungs after a maximum inspiration is the **total lung capacity** (TLC), and that after a maximum expiration is the **residual volume** (RV).

The **functional residual capacity** (FRC) is the volume of the lungs at the end of a normal breath, when the respiratory muscles are relaxed. It is determined by the balance between *outward elastic recoil* of the chest wall and *inward elastic recoil* of the lungs. These are coupled by the fluid in the small pleural space, which therefore has a negative pressure (**intrapleural pressure**: –0.2 to –0.5 kPa). Perforation of the chest therefore allows air to be sucked into the pleural space, and the chest wall expands while the lung collapses (**pneumothorax**). Diseases that affect lung elastic recoil alter FRC; *fibrosis* increases recoil and therefore reduces FRC, whereas in *emphysema*, where lung structure is lost, recoil is reduced and FRC increases.

During **inspiration**, the expansion of the thoracic cavity makes the intrapleural pressure more negative, causing the lungs and alveoli to expand, and reducing the alveolar pressure. This creates a pressure gradient between the alveoli and the mouth, drawing air into the lungs. During **expiration**, intrapleural and alveolar pressures rise, although, except during forced expiration (e.g. coughing), the intrapleural pressure remains negative throughout the cycle because expiration is normally *passive*.

The **dead space** refers to the volume of the airways that does not take part in gas exchange. The **anatomical dead space** includes the respiratory tract down to the terminal bronchioles; it is normally ~150 mL. The **alveolar dead space** refers to alveoli incapable of gas exchange; in health, it is negligible. The **physiological dead space** is the sum of the anatomical and alveolar dead spaces.

26 Lung mechanics

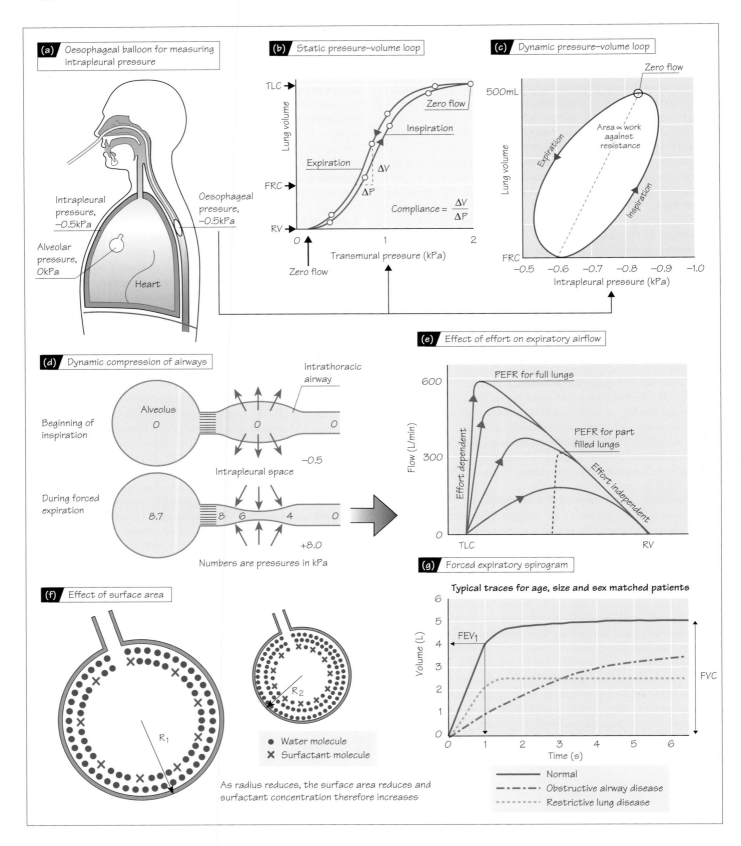

(a) Oesophageal balloon for measuring intrapleural pressure

Intrapleural pressure, −0.5kPa

Oesophageal pressure, −0.5kPa

Alveolar pressure, 0kPa

Heart

(b) Static pressure–volume loop

TLC

Zero flow

Inspiration

Lung volume

FRC

Expiration

ΔV

ΔP

RV

$$\text{Compliance} = \frac{\Delta V}{\Delta P}$$

0   1   2

Transmural pressure (kPa)

Zero flow

(c) Dynamic pressure–volume loop

Zero flow

500mL

Area ∝ work against resistance

Lung volume

Expiration

Inspiration

FRC

−0.5 −0.6 −0.7 −0.8 −0.9 −1.0

Intrapleural pressure (kPa)

(d) Dynamic compression of airways

Intrathoracic airway

Beginning of inspiration

Alveolus

0

0   0

Intrapleural space

−0.5

During forced expiration

8.7

8 6   4   0

+8.0

Numbers are pressures in kPa

(e) Effect of effort on expiratory airflow

PEFR for full lungs

600

PEFR for part filled lungs

Flow (L/min)

Effort dependent

300

Effort independent

0

TLC   RV

(f) Effect of surface area

R_1

R_2

● Water molecule
✗ Surfactant molecule

As radius reduces, the surface area reduces and surfactant concentration therefore increases

(g) Forced expiratory spirogram

Typical traces for age, size and sex matched patients

6

5

4

FEV_1

Volume (L)

3

2

FVC

1

0

0   1   2   3   4   5   6

Time (s)

——— Normal
—·—·— Obstructive airway disease
········· Restrictive lung disease

Physiology at a Glance, Third Edition. Jeremy P.T. Ward and Roger W.A. Linden.

 © 2013 John Wiley & Sons, Ltd. Published 2013 by John Wiley & Sons, Ltd.

The respiratory muscles have to overcome resisting forces during breathing. These are primarily the **elastic resistance** in the chest wall and lungs, and the resistance to air flow (**airway resistance**).

Lung compliance

The **static compliance** ('stretchiness') of the lungs (C_L) is defined as the change in volume per unit change in distending pressure ($C_L = \Delta V/\Delta P$) when there is no air flow. The distending pressure is the **transmural** (alveolar–intrapleural) pressure (Chapter 25). The intrapleural pressure can be measured with an oesophageal balloon (Fig. 26a). The alveolar pressure is the same as the mouth pressure (i.e. zero) if no air is flowing. The subject breathes in steps and the intrapleural pressure is measured at each held volume. A typical static **pressure–volume plot** is shown in Fig. 26b. The inspiratory and expiratory curves are slightly different (*hysteresis*), typical for elastic systems. The **static lung compliance** is the maximum slope, generally just above the functional residual capacity (FRC), and is normally ~1.5 L/kPa, although this is dependent on age, size and sex. The static compliance is reduced by lung *fibrosis* (stiffer lungs).

The **dynamic compliance** is measured during continuous breathing, and therefore includes a component due to **airway resistance**. The dynamic pressure–volume loop (Fig. 26c) has a point at each end where the flow is zero; the slope of the line between these points is the **dynamic compliance**. This is normally similar to the static compliance, but can be altered in disease. The width of the curve reflects the pressure required to suck in or expel air; the *area of the curve* is therefore a measure of the work done against airway resistance.

Surfactant and the alveolar air–fluid interface

The **surface tension** of the fluid lining the alveoli contributes to lung stiffness, as the attraction of water molecules at the air–fluid interface tends to collapse the alveoli. This is a manifestation of **Laplace's law** (Chapter 11), which shows that the pressure in a bubble (or alveolus) is proportional to the surface tension (T) and radius ($P \propto T/r$). A small bubble will therefore have a higher pressure than a larger one and, if connected, will collapse into it. The inward force created by this surface tension also tends to suck fluid into the alveoli (**transudation**). In the lung, these problems are minimized by **surfactant**, secreted by **type II pneumocytes** (Chapter 25). Surfactant is a mixture of **phospholipids** that floats on the alveolar fluid surface, and reduces surface tension. As the alveoli shrink, the effective concentration of surfactant increases, further lowering the surface tension (Fig. 26f). This more than balances the effect of reducing radius (as r falls, so does T). Surfactant also reduces lung stiffness and transudation. Premature babies may not have sufficient surfactant and develop **neonatal respiratory distress syndrome**, with stiff lungs, lung collapse and transudation.

Airway resistance

Flow through the airways is described by **Darcy's law**; flow = $(P_1 - P_2)/R$ (Chapter 11), where P_1 is the alveolar pressure, P_2 is the mouth pres-

sure and R is the resistance to air flow. The airway resistance is determined by the airway radius, according to **Poiseuille's law**, and whether the flow is laminar or turbulent (Chapter 11).

The airway resistance is increased by factors that constrict the airway smooth muscle (**bronchoconstrictors**). These include the reflex release of **muscarinic** neurotransmitters from parasympathetic nerve endings, generally due to the activation of **irritant receptors** (Chapter 29), and numerous mediators released by inflammatory cells (e.g. **histamine, prostaglandins, leukotrienes**), e.g. in **asthma**. Increased mucus production also narrows the lumen and increases the resistance. Sympathetic stimulation, adrenaline (epinephrine) and salbutamol cause relaxation and **bronchodilatation** via β_2-adrenoceptors on the smooth muscle.

Effect of transmural pressure. Expiration is normally passive (Chapter 29). Forced expiration increases the intrapleural and thus alveolar pressure, increasing the pressure gradient to the mouth and therefore theoretically leading to increased flow. However, although expiration from fully inflated lungs is indeed **effort dependent**, towards the end of the breath, increasing force does not increase flow, i.e. it is **effort independent** (Fig. 26e). This occurs as a result of the pressure gradient between the alveoli and the mouth. Midway between them, generally in the bronchi, the pressure in the airway falls below the intrapleural pressure, causing the airway to collapse (**dynamic compression**; Fig. 26d). As there is now no flow, the pressure rises again until it is greater than the intrapleural pressure, and the airway re-opens. This sequence happens repeatedly, producing the brassy sound heard during forced expiration. This does not occur in normal expiration because the intrapleural pressure remains negative throughout. In diseases in which the airways are already narrowed (e.g. *asthma*), this leads to expiratory wheezing and air trapping.

Lung function tests

Lung volumes can be measured using a simple spirometer (Chapter 25). Airway resistance and lung compliance can be assessed indirectly by measuring the forced expiratory flows and volumes. The easiest and quickest measurement is the **peak expiratory flow rate** (PEFR). PEFR is decreased if the airway resistance is increased (**obstructive disease**), and is commonly used to follow an already diagnosed condition, e.g. asthma. It is, however, dependent on the initial lung volume (Fig. 26e). Plots of the **forced expiratory volume against time** provide more information. Subjects breathe out from total lung capacity to residual volume as fast as possible; this is the **forced vital capacity** (FVC), and a typical trace is shown in Figure 26g. The forced expiratory volume in 1 s (**FEV$_1$**) reflects the airway resistance; it is normally expressed as a ratio to FVC (**FEV$_1$/FVC**) to correct for lung volume, and is usually 0.75–0.90. It can be used to distinguish between **obstructive** (increased airway resistance) and **restrictive** (decreased lung compliance) diseases. In asthma, for example, FEV$_1$/FVC is typically <0.7. In restrictive disease (e.g. lung fibrosis), FEV$_1$ and FVC are low, but FEV$_1$/FVC is normal or even increased due to greater elastic recoil (Fig. 26g).

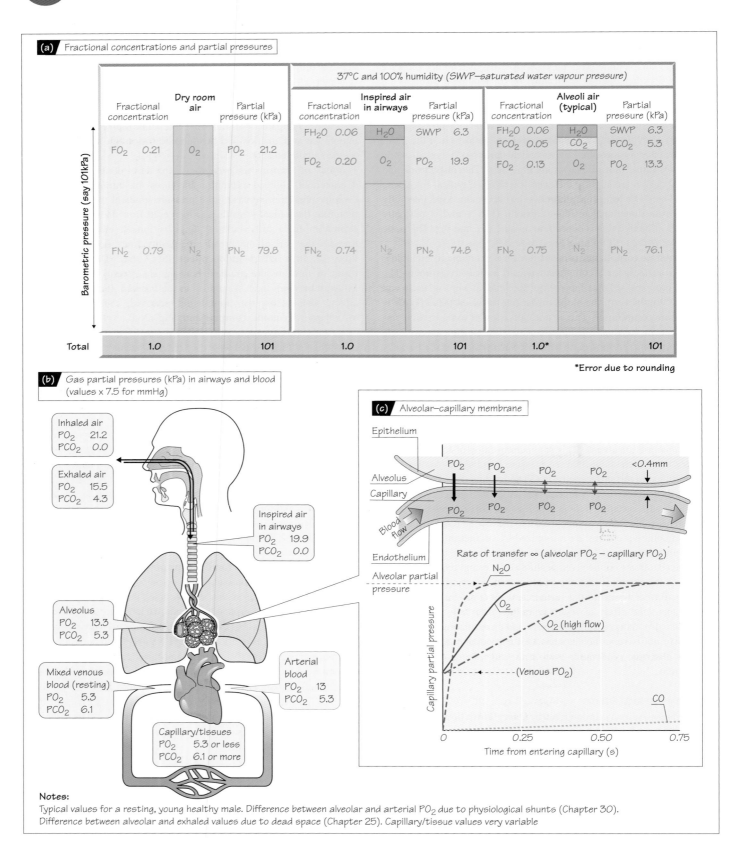

Notes:

Typical values for a resting, young healthy male. Difference between alveolar and arterial PO_2 due to physiological shunts (Chapter 30). Difference between alveolar and exhaled values due to dead space (Chapter 25). Capillary/tissue values very variable

Physiology at a Glance, Third Edition. Jeremy P.T. Ward and Roger W.A. Linden.

64 © 2013 John Wiley & Sons, Ltd. Published 2013 by John Wiley & Sons, Ltd.

Dry air contains 78.1% N_2 and 21% O_2; other inert gases account for the balance (0.9%), but are normally pooled with N_2 (i.e. N_2 = 79%). The small amount of CO_2 in air (<0.04%) is usually ignored.

Partial pressures and fractional concentrations (Fig. 27a)

The volume of a fixed amount of gas is inversely proportional to the pressure (V ∝ 1/P; **Boyle's law**) and proportional to the absolute temperature (V ∝ T; **Charles' law**). An ideal gas occupies 22.4 L per mole at 1 atm (101 kPa, 760 mmHg) and 0 °C (273 K), and thus the volume of each gas in a mixture is directly proportional to the quantity of that gas in moles. The term **fractional concentration (F)** can therefore be used to denote the relative quantities of gases in any mixture; thus F_{N_2} is 0.79 in *dry* air, and F_{O_2} is 0.21. The **partial pressure** of each gas in a mixture is that part of the total (e.g. barometric) pressure that is exerted by that gas, and is directly proportional to the quantity. Thus, according to **Dalton's law**, the partial pressure of O_2 (P_{O_2}) in dry air is F_{O_2} × barometric pressure (P_B), e.g. 0.21 × 101 kPa × 21.2 kPa. At the summit of Everest, P_B is ~34 kPa, but the relative proportions of gases are the same as at sea level, and so P_{O_2} is 0.21 × 34 kPa = 7.14 kPa.

Water vapour pressure. Water vapour behaves like any other gas, and exerts a partial pressure. The maximum or **saturated water vapour pressure (SWVP)** depends on the temperature: 2.33 kPa at 20 °C and 6.3 kPa at 37 °C. Inspired air quickly reaches body temperature and becomes fully humidified (100% saturated) in the airways. Water vapour dilutes the other gases, so that P_{N_2} and P_{O_2} will be lower than in dry air. Thus, P_{O_2} will be 0.21 × (P_B – saturated water vapour pressure) or, under these conditions, 0.21 × (101 – 6.3) = 19.9 kPa (Fig. 27a). The water vapour content of room air depends on the conditions (e.g. desert vs seaside); 40% humidity denotes 40% of the predicted SWVP for that temperature.

Standardization. From the above and Boyle's and Charles' laws, it should be clear that gas volumes and partial pressures cannot be compared unless corrected to a standardized pressure, temperature and humidity. Two standards are commonly used: standard temperature and pressure, dry gas (**STPD**), corrected to 1 standard atm (10 kPa), 0 °C and dry gas; and body temperature and pressure (1 atm), saturated with water (**BTPS**).

Gases dissolved in body fluids

The quantity of gas dissolving in a fluid is described by **Henry's law**: dissolved gas concentration = partial pressure of gas above fluid × **solubility** of that gas in that fluid. The solubility tends to decrease with a rise in temperature, and varies significantly between gases. For example, CO_2 is 20 times more soluble than O_2 in water, so that water exposed to the same partial pressures of CO_2 and O_2 will contain 20 times as much CO_2 as O_2. Henry's law describes an *equilibrium* –

increasing the partial pressure of a gas will cause more to dissolve in the fluid until a new equilibrium is reached. The concept of a partial pressure of gas dissolved in a fluid (e.g. P_{O_2} of blood) is sometimes difficult to understand, but merely reflects the partial pressure that would be required to dissolve that amount of gas in the fluid, according to Henry's law. From the above, it can be deduced that the movement of gases between gas and fluid phases (e.g. alveolar air and capillary blood) will be dependent on the **difference in partial pressures** rather than the concentration. Typical values for partial pressures in the airways and blood are shown in Figure 27b.

Diffusion across the alveolar–capillary membrane (Fig. 27c)

Diffusion is discussed in Chapter 11. The rate of gas flow across the alveolar–capillary membrane = permeability × area × (difference in partial pressures), where the permeability depends on the membrane thickness, gas molecular weight and its solubility in the membrane (Chapter 11). Although CO_2 is larger than O_2, it crosses the membrane faster because it is more soluble in biological membranes. For gas transfer across the lungs, the permeability and area are commonly combined as the **diffusing capacity (D_L)** for that gas, a measure of alveolar–capillary membrane function. Thus, the rate of O_2 transfer = $D_L O_2$ × (alveolar P_{O_2} – lung capillary P_{O_2}), or $D_L O_2$ = O_2 **uptake from lungs/(alveolar P_{O_2} – lung capillary P_{O_2})**. $D_L O_2$ is sometimes called the **transfer factor**. $D_L O_2$ cannot be estimated directly, because capillary P_{O_2} cannot be measured. However, the factors affecting O_2 diffusion also affect carbon monoxide (CO) diffusion. CO binds extremely strongly to haemoglobin, and so, if low concentrations of CO are inhaled, CO diffusing into the blood is completely bound to haemoglobin and capillary P_{CO} remains close to zero (Fig. 27c). Thus, $D_L CO$ = CO uptake from lungs/alveolar P_{CO}, and can be easily measured as an estimate of alveolar–capillary transfer function. $D_L CO$ is reduced by a decrease in lung exchange area (e.g. emphysema) or an increase in alveolar–capillary membrane thickness (e.g. lung fibrosis, oedema).

Diffusion and perfusion limitation (Fig. 27c)

Because CO binds so avidly and rapidly to haemoglobin, at low concentrations its rate of transfer into the blood is not affected by the blood flow, because there is always plenty of haemoglobin, it is limited solely by its rate of diffusion across the alveolar–capillary membrane, i.e. transfer is **diffusion limited**. For a poorly soluble gas, however (e.g. the anaesthetic nitrous oxide, N_2O), the partial pressure in the blood rapidly reaches equilibrium with alveolar air, preventing further diffusion. In this case, increased blood flow will increase the rate of transfer, i.e. transfer is **perfusion limited**. O_2 transfer is normally perfusion limited.

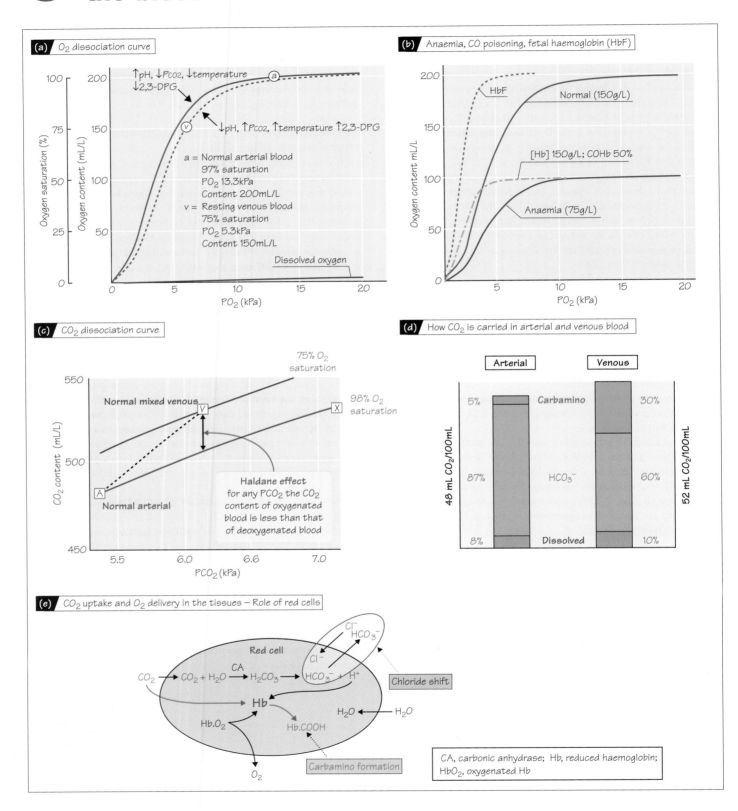

(a) O_2 dissociation curve

↑pH, ↓P_{CO2}, ↓temperature
↓2,3-DPG

↓pH, ↑P_{CO2}, ↑temperature ↑2,3-DPG

a = Normal arterial blood
97% saturation
PO_2 13.3kPa
Content 200mL/L
v = Resting venous blood
75% saturation
PO_2 5.3kPa
Content 150mL/L

Dissolved oxygen

Oxygen saturation (%)
Oxygen content (mL/L)
PO_2 (kPa)

(b) Anaemia, CO poisoning, fetal haemoglobin (HbF)

HbF
Normal (150g/L)
[Hb] 150g/L; COHb 50%
Anaemia (75g/L)

Oxygen content mL/L
PO_2 (kPa)

(c) CO_2 dissociation curve

75% O_2 saturation
98% O_2 saturation

Normal mixed venous
V
X
A
Normal arterial

Haldane effect
for any PCO_2 the CO_2 content of oxygenated blood is less than that of deoxygenated blood

CO_2 content (mL/L)
PCO_2 (kPa)

(d) How CO_2 is carried in arterial and venous blood

Arterial		Venous
5%	Carbamino	30%
87%	HCO_3^-	60%
8%	Dissolved	10%

48 mL CO_2/100mL
52 mL CO_2/100mL

(e) CO_2 uptake and O_2 delivery in the tissues – Role of red cells

Red cell

CO_2 → $CO_2 + H_2O$ →[CA]→ H_2CO_3 → $HCO_3^- + H^+$

Cl^-
HCO_3^-
Cl^-
Chloride shift

Hb

Hb.O_2
Hb.COOH

H_2O ← H_2O

Carbamino formation

O_2

CA, carbonic anhydrase; Hb, reduced haemoglobin; HbO2, oxygenated Hb

Physiology at a Glance, Third Edition. Jeremy P.T. Ward and Roger W.A. Linden.
 © 2013 John Wiley & Sons, Ltd. Published 2013 by John Wiley & Sons, Ltd.

Oxygen

The resting O_2 consumption in adults is ~250 mL/min, rising to >4000 mL/min during heavy exercise. The O_2 **solubility** in plasma is, however, low and at a Po_2 of 13 kPa blood contains only 3 mL/L of dissolved O_2 in solution. Most O_2 is therefore carried bound to **haemoglobin** in red blood cells. Each gram of haemoglobin can combine with 1.34 mL of O_2 and so, for a haemoglobin concentration [Hb] of 150 g/L, blood can contain a maximum of 200 mL/L of O_2 (**O_2 capacity**). The actual amount of O_2 bound to haemoglobin (**O_2 content**) depends on the Po_2, and the **percentage O_2 saturation** = content/capacity × 100 (Fig. 28a). Each haemoglobin molecule binds up to four O_2 molecules; binding is **cooperative**, so that the binding of each O_2 molecule makes it easier for the next. This steepens the **O_2 haemoglobin dissociation curve**, which describes the relationship between blood O_2 content and Po_2 (Fig. 28a). The curve flattens above ~8 kPa Po_2 as all binding sites become occupied. Thus, for a normal arterial Po_2 (~13 kPa) and [Hb], the blood is ~97% saturated and contains slightly less than 200 mL/L of O_2. Because the dissociation curve is flat in this region, any increase in Po_2 (breathing O_2-enriched air) will have little effect on content. On the steep part of the curve, however (<8 kPa Po_2), small changes in Po_2 will have large effects on content.

Oxygen uptake and delivery. The high Po_2 in the lungs facilitates O_2 binding to haemoglobin, whereas the low Po_2 in the tissues encourages release. The dissociation curve is shifted to the right (reduced affinity, facilitating O_2 release) by a fall in pH, a rise in Pco_2 (**Bohr shift**) and an increase in temperature, which occur in active tissues (Fig. 28a). The metabolic by-product **2,3-diphosphoglycerate** (2,3-DPG) also causes a right shift. In the lungs, Pco_2 falls, the pH consequently rises and the temperature is reduced; these all increase affinity and shift the curve to the left, facilitating O_2 uptake.

Anaemia. This is an abnormally low [Hb]; the O_2 capacity is therefore less and the O_2 content at any Po_2 is reduced (Fig. 28b). Arterial Po_2 and O_2 saturation remain normal. In order to deliver the same amount of O_2 to the tissues, the capillary Po_2 would have to fall further than normal (Fig. 28b), reducing the driving force for O_2 diffusion into the tissues. The latter may become inadequate for metabolism, especially during exercise, although a 50% reduction in [Hb] does not usually cause symptoms at rest.

Carbon monoxide. Carbon monoxide (CO) binds 240 times more strongly than O_2 to haemoglobin and, by occupying O_2-binding sites, reduces the O_2 capacity. However, unlike anaemia, CO also increases the affinity and shifts the dissociation curve to the left, making O_2 release to the tissues more difficult. Thus, if 50% of haemoglobin is bound to CO, Po_2 needs to fall much further than in anaemia to release the same amount of O_2, causing symptoms of severe hypoxia (headache, convulsions, coma, death) (Fig. 28b).

Fetal haemoglobin. Fetal haemoglobin (HbF) binds 2,3-DPG less strongly than does adult haemoglobin (HbA), and so the dissociation curve is shifted to the left. This facilitates the transfer of O_2 from maternal blood to the fetus, where the arterial Po_2 is only ~5 kPa (Fig. 28b).

Carbon dioxide

CO_2 is formed in the tissues and transported to the lungs where it is expired. Blood can carry much more CO_2 than O_2, as can be seen in the **CO_2 dissociation curve** (Fig. 28c). This is also more linear than the O_2 dissociation curve and does not plateau. CO_2 is transported as **bicarbonate, carbamino compounds** and simply **dissolved** in plasma (Fig. 28d).

Bicarbonate. Approximately 60% of CO_2 is carried as bicarbonate. Water and CO_2 combine to form carbonic acid (H_2CO_3) and thence bicarbonate (HCO_3^-): $CO_2 + H_2O \Leftrightarrow H_2CO_3 \Leftrightarrow HCO_3^- + H^+$. The left side of the equation is normally slow, but speeds up dramatically in the presence of **carbonic anhydrase**, found in red cells. Bicarbonate is therefore formed preferentially in red cells, from which it easily diffuses out. Red cells are, however, impermeable to H^+ ions, and Cl^- enters the cell to maintain electrical neutrality (**chloride shift**) (Fig. 28e). H^+ binds avidly to **deoxygenated** (*reduced*) haemoglobin (haemoglobin acts as a buffer), and so there is little increase in [H^+] to impede further bicarbonate formation. *Oxygenated* haemoglobin does not bind H^+ as well, and so in the lungs H^+ dissociates from haemoglobin and shifts the CO_2–HCO_3^- equation to the left, assisting CO_2 unloading from the blood (Fig. 28e); the reverse occurs in the tissues. This contributes to the **Haldane effect**, which states that, for any Pco_2, the CO_2 content of oxygenated blood is less than that of deoxygenated blood. Thus the red line A–X in Figure 28c shows the relationship between CO_2 content and Pco_2 if the blood remained 98% saturated with O_2. Mixed venous O_2 saturation is however ~75%, so as the blood becomes oxygenated in the lungs or deoxygenated in the tissues, the relationship between CO_2 content and Pco_2 actually follows the dashed line A–V.

Carbamino compounds. These compounds are formed by the reaction of CO_2 with protein amino groups: CO_2 + protein-NH_2 $\Leftrightarrow$ protein-NHCOOH. The most prevalent protein in blood is haemoglobin, which forms **carbaminohaemoglobin** with CO_2. This occurs more readily for deoxygenated than oxygenated haemoglobin, contributing to the Haldane effect (Fig. 28c). Carbamino compounds account for 30% of CO_2 carriage.

Dissolved carbon dioxide. CO_2 is 20 times more soluble than O_2 in plasma, and ~10% of CO_2 in blood is carried in **solution**.

Hyperventilation and hypoventilation

Doubling the rate of ventilation halves the alveolar and arterial Pco_2. Ventilation is normally closely matched to the metabolic rate as reflected by CO_2 production (Chapter 29). **Hyperventilation** (overventilation) and **hypoventilation** (underventilation) are defined in terms of arterial Pco_2, so that a subject is *hyperventilating* when Pco_2 is <5.3 kPa, and hypoventilating when Pco_2 is >5.9 kPa. Rapid breathing in exercise is *not* hyperventilation, as this is appropriate for increased CO_2 production and Pco_2 does not fall. Hyperventilation cannot normally increase the O_2 content, as arterial haemoglobin is already nearly fully saturated. The fall in Pco_2 (**hypocapnia**) during hyperventilation causes light-headedness, visual disturbances due to cerebral vasoconstriction (Chapter 24) and muscle cramps (tetany). Hyperventilation can be caused by pain, hysteria and strong emotion. Hypoventilation causes a high Pco_2 (**hypercapnia**) and a low Po_2 (**hypoxia**), and may be caused by head injury or respiratory disease.

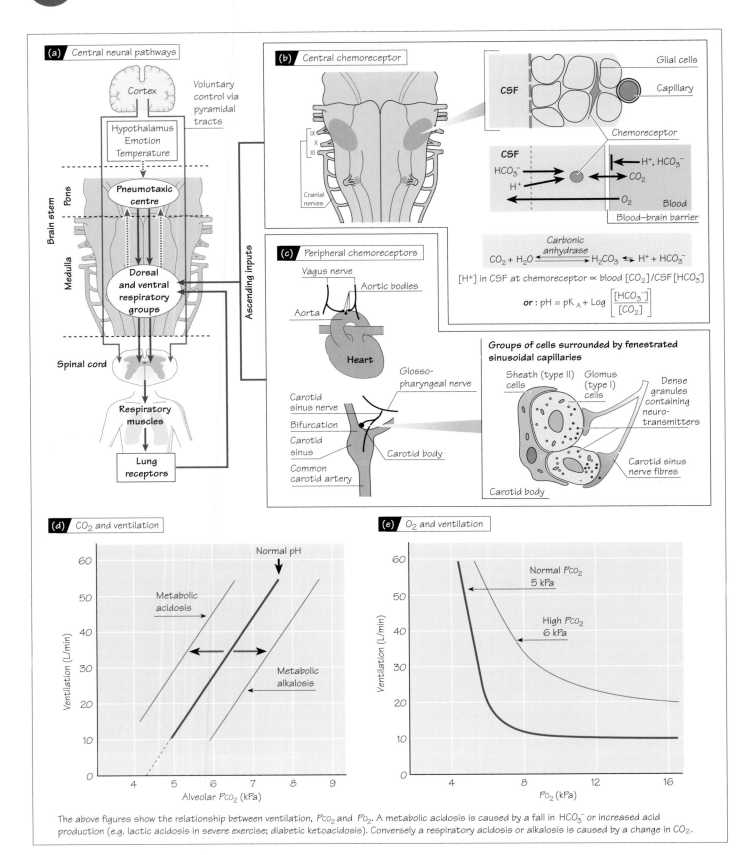

(a) Central neural pathways

Cortex

Voluntary control via pyramidal tracts

Hypothalamus
Emotion
Temperature

Pneumotaxic centre

Dorsal and ventral respiratory groups

Brain stem — Pons — Medulla

Spinal cord

Respiratory muscles

Lung receptors

Ascending inputs

(b) Central chemoreceptor

ix
x
xi

Cranial nerves

Glial cells

Capillary

CSF

Chemoreceptor

CSF

HCO_3^-

H^+

H^+, HCO_3^-

CO_2

O_2 Blood

Blood–brain barrier

Carbonic anhydrase

$$CO_2 + H_2O \rightleftharpoons H_2CO_3 \rightleftharpoons H^+ + HCO_3^-$$

$[H^+]$ in CSF at chemoreceptor $\propto$ blood $[CO_2]/$CSF $[HCO_3^-]$

or : $pH = pK_A + Log \dfrac{[HCO_3^-]}{[CO_2]}$

(c) Peripheral chemoreceptors

Vagus nerve

Aortic bodies

Aorta

Heart

Glosso-pharyngeal nerve

Carotid sinus nerve

Bifurcation

Carotid sinus

Common carotid artery

Carotid body

Groups of cells surrounded by fenestrated sinusoidal capillaries

Sheath (type II) cells

Glomus (type I) cells

Dense granules containing neuro-transmitters

Carotid sinus nerve fibres

Carotid body

(d) CO_2 and ventilation

Normal pH

Metabolic acidosis

Metabolic alkalosis

Ventilation (L/min): 60, 50, 40, 30, 20, 10, 0

Alveolar P_{CO_2} (kPa): 4, 5, 6, 7, 8, 9

(e) O_2 and ventilation

Normal P_{CO_2} 5 kPa

High P_{CO_2} 6 kPa

Ventilation (L/min): 60, 50, 40, 30, 20, 10, 0

P_{O_2} (kPa): 4, 8, 12, 16

The above figures show the relationship between ventilation, P_{CO_2} and P_{O_2}. A metabolic acidosis is caused by a fall in HCO_3^- or increased acid production (e.g. lactic acidosis in severe exercise; diabetic ketoacidosis). Conversely a respiratory acidosis or alkalosis is caused by a change in CO_2.

Physiology at a Glance, Third Edition. Jeremy P.T. Ward and Roger W.A. Linden.

 © 2013 John Wiley & Sons, Ltd. Published 2013 by John Wiley & Sons, Ltd.

Ventilation of the lungs provides O_2 for the tissues and removes CO_2. Breathing must therefore be closely matched to metabolism for adequate O_2 delivery and to prevent a build-up of CO_2. A **central pattern generator** located in the brain stem sets the basic rhythm and pattern of ventilation and controls the respiratory muscles. It is modulated by higher centres and feedback from **sensors**, including **chemoreceptors**, and **lung mechanoreceptors** (Fig. 29a). The neural networks are complex, as breathing must be coordinated with coughing, swallowing and speech, and are not fully understood.

The brain stem and central pattern generator

The brain stem includes diffuse groups of respiratory neurones in the **pons** and **medulla** that act together as the central pattern generator (Fig. 29a); it is unclear whether there is a single pacemaker region. Some neurones only show activity during inspiration or expiration, and these exhibit **reciprocal inhibition**, i.e. inspiration inhibits expiration and *vice versa*. The medulla contains **dorsal** and **ventral respiratory groups** that receive input from the chemoreceptors and lung receptors and drive the respiratory muscle motor neurones [intercostals, phrenic (diaphragm), abdominal]. The medullary respiratory groups also provide ascending input to and receive descending input from the **pneumotaxic centre** in the pons, which is critical for normal breathing. The pneumotaxic centre receives input from the hypothalamus and higher centres, coordinates medullary homeostatic functions with factors such as emotion and temperature, and affects the pattern of breathing. Voluntary control is mediated by cortical motor neurones in the **pyramidal tract**, which by-passes the respiratory neurones in the brainstem.

Chemoreception

Chemoreceptors detect arterial P_{CO_2}, P_{O_2} and pH – P_{CO_2} is the most important. Alveolar P_{CO_2} ($P_{A_{CO_2}}$) is normally ~5.3 kPa (40 mmHg), and $P_{A_{O_2}}$ normally 13 kPa (100 mmHg). An increase in $P_{A_{CO_2}}$ causes ventilation to rise in an almost linear fashion (Fig. 29d). Increased acidity of the blood (e.g. lactic acidosis in severe exercise) causes the relationship between P_{CO_2} and ventilation to shift to the left, and decreased acidity causes a shift to the right. Conversely, P_{O_2} normally stimulates ventilation only when it falls below ~8 kPa (~60 mmHg) (Fig. 29e). However, when a fall in P_{O_2} is accompanied by an increase in P_{CO_2}, the resultant increase in ventilation is greater than would be expected from the effects of either alone; there is thus a **synergistic** (more than additive) relationship between P_{O_2} and P_{CO_2} (Fig. 29e).

The **central chemoreceptor** comprises a collection of neurones near the ventrolateral surface of the medulla, close to the exit of the cranial nerves IX and X (Fig. 29b). It responds *indirectly* to blood P_{CO_2}, but does **not** respond to changes in P_{O_2}. Although CO_2 can easily diffuse across the **blood–brain barrier** from the blood into the cerebrospinal fluid (CSF), H^+ and HCO_3^- cannot. As a result, the pH of the CSF around the chemoreceptor is determined by the arterial P_{CO_2} and CSF HCO_3^-, according to the Henderson–Hasselbalch equation (Fig. 29b). A rise in blood P_{CO_2} therefore makes the CSF more acid; this is detected by the chemoreceptor, which increases ventilation to blow off CO_2. The central chemoreceptor is responsible for ~80% of the response to CO_2 in humans. Its response is delayed because CO_2 has to diffuse across the blood–brain barrier. As the blood–brain barrier is impermeable to H^+, the central chemoreceptor is not affected by blood pH.

The **peripheral chemoreceptors** are located in the carotid and aortic bodies (Fig. 29c). The **carotid bodies** are small distinct structures located at the bifurcation of the common carotid arteries, and are innervated by the carotid sinus nerve and thence the glossopharyngeal nerve. The carotid body is formed from **glomus** (type I) and **sheath** (type II) cells. Glomus cells are chemoreceptive, contain neurotransmission-rich dense granules and contact carotid sinus nerve axons. The **aortic bodies** are located on the aortic arch and are innervated by the vagus. They are similar to carotid bodies but functionally less important. Peripheral chemoreceptors respond to changes in P_{CO_2}, H^+ and, importantly, P_{O_2}. They are responsible for ~20% of the response to increased P_{CO_2}.

Lung receptors

Various types of lung receptor provide feedback from the lungs to the respiratory centre. In addition, **pain** often causes brief apnoea (cessation of breathing) followed by rapid breathing, and **mechanical** or **noxious** stimulation of receptors in the trigeminal region and larynx causes apnoea or spasm of the larynx.

Stretch receptors. These are located in the bronchial walls. Stimulation (by stretch) causes short, shallow breaths, and delay of the next inspiratory cycle. They provide negative feedback to turn off inspiration. They are mostly **slowly adapting** (continue to fire with sustained stimulation) and are innervated by the vagus. They are largely responsible for the **Hering–Breuer inspiratory reflex**, in which lung inflation inhibits inspiration to prevent overinflation.

Juxtapulmonary (J) receptors. These are located on the alveolar and bronchial walls close to the capillaries. They cause depression of somatic and visceral activity by producing rapid shallow breathing or apnoea, a fall in heart rate and blood pressure, laryngeal constriction and relaxation of the skeletal muscles via spinal neurones. They are stimulated by increased alveolar wall fluid, oedema, microembolisms and inflammation. The afferent nerves are small unmyelinated (C-fibre) or myelinated nerves in the vagus.

Irritant receptors. These are located throughout the airways between epithelial cells. In the trachea they cause cough, and in the lower airways hyperpnoea (rapid breathing); stimulation also causes bronchial and laryngeal constriction. They are also responsible for the deep augmented breaths every 5–20 min at rest, reversing the slow collapse of the lungs that occurs in quiet breathing, and may be involved in the first deep gasps of the newborn. They are stimulated by irritant gases, smoke and dust, rapid large inflations and deflations, airway deformation, pulmonary congestion and inflammation. The afferent nerves are rapidly adapting myelinated fibres in the vagus.

Proprioceptors (position/length sensors). These are located in the Golgi tendon organs, muscle spindles and joints. They are important for matching increased load, and maintaining optimal tidal volume and frequency. They are stimulated by shortening and load in the respiratory muscles (but not the diaphragm). Afferents run to the spinal cord via the dorsal roots. It should be noted that input from non-respiratory muscles and joints can also stimulate breathing.

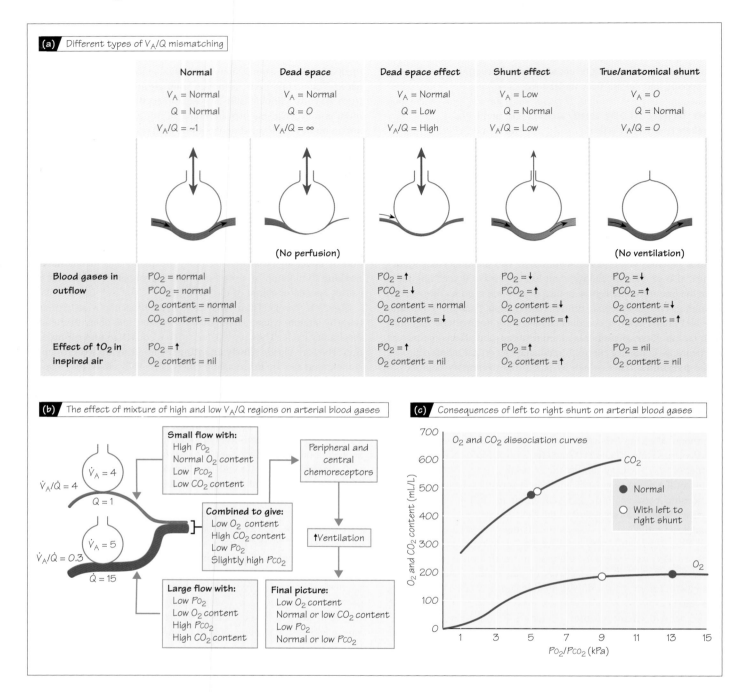

(a) Different types of V_A/Q mismatching

	Normal	Dead space	Dead space effect	Shunt effect	True/anatomical shunt
	V_A = Normal	V_A = Normal	V_A = Normal	V_A = Low	V_A = 0
	Q = Normal	Q = 0	Q = Low	Q = Normal	Q = Normal
	V_A/Q = ~1	V_A/Q = ∞	V_A/Q = High	V_A/Q = Low	V_A/Q = 0
		(No perfusion)			(No ventilation)
Blood gases in outflow	PO_2 = normal PCO_2 = normal O_2 content = normal CO_2 content = normal		PO_2 = ↑ PCO_2 = ↓ O_2 content = normal CO_2 content = ↓	PO_2 = ↓ PCO_2 = ↑ O_2 content = ↓ CO_2 content = ↑	PO_2 = ↓ PCO_2 = ↑ O_2 content = ↓ CO_2 content = ↑
Effect of ↑O_2 in inspired air	PO_2 = ↑ O_2 content = nil		PO_2 = ↑ O_2 content = nil	PO_2 = ↑ O_2 content = ↑	PO_2 = nil O_2 content = nil

(b) The effect of mixture of high and low V_A/Q regions on arterial blood gases

$\dot{V}_A/\dot{Q} = 4$
$\dot{V}_A = 4$
Q = 1

$\dot{V}_A/\dot{Q} = 0.3$
$\dot{V}_A = 5$
$\dot{Q} = 15$

Small flow with:
High PO_2
Normal O_2 content
Low PCO_2
Low CO_2 content

Large flow with:
Low PO_2
Low O_2 content
High PCO_2
High CO_2 content

Combined to give:
Low O_2 content
High CO_2 content
Low PO_2
Slightly high PCO_2

Peripheral and central chemoreceptors

↑Ventilation

Final picture:
Low O_2 content
Normal or low CO_2 content
Low PO_2
Normal or low PCO_2

(c) Consequences of left to right shunt on arterial blood gases

O_2 and CO_2 dissociation curves

O_2 and CO_2 content (mL/L)

CO_2

● Normal
○ With left to right shunt

O_2

PO_2/PCO_2 (kPa)

Physiology at a Glance, Third Edition. Jeremy P.T. Ward and Roger W.A. Linden.
 © 2013 John Wiley & Sons, Ltd. Published 2013 by John Wiley & Sons, Ltd.

Ventilation–perfusion matching (Fig. 30a)

At rest, total alveolar ventilation (V_A) is similar to total pulmonary capillary perfusion (Q), or about 5 L/min. For optimal gas exchange, all regions of the lung should ideally have a **ventilation–perfusion ratio** (V_A/Q) of unity. When there are variations significantly away from unity, either lower or higher, this is referred to as **ventilation–perfusion mismatch**. In a **right to left shunt** (see below), for example, ventilation is zero and $V_A/Q = \infty$; whereas when an embolism blocks a pulmonary artery, perfusion in the part of the lung fed by that artery is zero, and $V_A/Q = 0$. Regions of the lung that have a V_A/Q value much greater than unity have excessive ventilation, and blood derived from them will have a high Po_2 and a low Pco_2 (**dead space effect**). Regions with a V_A/Q value much less than unity have a **shunt** or **venous admixture** effect; there is some gas exchange, but the blood has a lower than normal Po_2 and a higher than normal Pco_2.

Effect of ventilation–perfusion mismatch on arterial gases. Regions of high V_A/Q cannot compensate for regions of low V_A/Q. This is because of the way in which O_2 is carried in the blood (Chapter 28). Although regions of high V_A/Q produce blood with a high Po_2, this does not translate to any significant increase in O_2 content, as the haemoglobin in the blood is already close to saturation at the normal Po_2. Conversely, blood derived from regions with a low V_A/Q, especially if $Po_2 < 8$ kPa, will have a significantly reduced O_2 content (Fig. 30c). Regions of high V_A/Q are also most usually due to insufficient perfusion, so that the amount of blood, and therefore O_2, that such regions contribute to the total will be relatively small. As a result, the combined blood from regions with high and low V_A/Q will have a low O_2 content and a low Po_2, even if total ventilation and perfusion are matched for the whole lung.

The CO_2 content is less severely affected, because overventilated areas can lose extra CO_2 and partly compensate for underventilated areas. Moreover, a rise in Pco_2 will stimulate breathing via the chemoreceptors, allowing CO_2 to be corrected, or even overcorrected if Po_2 is sufficiently low (Chapter 29). Significant V_A/Q mismatching will therefore usually result in arterial blood with a low Po_2 but normal or low Pco_2 (Fig. 30b). Ventilation with O_2-enriched air will improve oxygenation in regions of low V_A/Q, but is not useful for shunts, as the enriched air never reaches the shunted blood. **Hypoxic pulmonary vasoconstriction** (Chapter 24) reduces the severity of V_A/Q mismatch by diverting blood from the affected region to well-ventilated areas.

Effect of gravity

The blood pressure at the base of the lungs is greater than that at the apex (top) because of the weight of the column of blood. This increased pressure distends the pulmonary blood vessels at the base and the flow is therefore increased. Conversely, blood flow at the apex may be reduced if the pulmonary venous pressure falls below the alveolar pressure, when the vessels will be compressed. The net result is that, on standing, pulmonary blood flow falls progressively on moving from the bottom of the lung to the top.

Gravity also affects the intrapleural pressure, which is thus less negative at the base of the lung than at the apex. Alveoli at the base are therefore less expanded at functional residual capacity, and thus have more potential for expansion during inspiration. As a result, ventilation is greatest at the base of the lung. Although the effects of gravity on perfusion and ventilation partly cancel each other out, ventilation is less affected than perfusion, so that V_A/Q is highest at the apex of the lung and lowest at the base. In the young, this relatively small variation has little effect on blood gases, but in the elderly, it may contribute to a low Po_2.

Right to left shunts

Part of the venous effluent of the bronchial and coronary circulations bypasses the lungs and enters the pulmonary vein and left ventricle, respectively (Chapter 16). Oxygenated blood from the lungs is therefore diluted by venous blood. These are **anatomical right to left shunts** and account for <2% of cardiac output in healthy individuals. Larger shunts can occur in disease when regions of the lung are not ventilated (e.g. lung collapse, pneumonia), or due to **congenital heart malformations**. When calculating the effects of right to left shunts on arterial blood, the blood **content** of O_2 and CO_2 needs to be considered. For a 20% shunt, the 80% of blood passing through the lungs will have normal arterial O_2 and CO_2 contents of 200 and 480 mL/L, respectively, whilst the 20% bypassing the lungs will have normal venous values of 150 and 520 mL/L, respectively. On combination, the blood will contain $(200 \times 0.8) + (150 \times 0.2) = 190$ mL/L of O_2, and $(480 \times 0.8) + (520 \times 0.2) = 488$ mL/L of CO_2. From the dissociation curves (Fig. 30c), it can be seen that this results in a fall in Po_2 from 13 to 9 kPa, whereas Pco_2 rises only marginally from 5.3 to 5.5 kPa because of the steeper curve. Changes in Pco_2 and Po_2 stimulate the chemoreceptors and increase ventilation (Chapter 29), so that arterial Pco_2 returns to normal. However, increased ventilation cannot increase blood O_2 content, as the haemoglobin of the blood passing through the lungs is already close to saturation. Thus, right to left shunts commonly result in a *low arterial Po_2* but a *normal or low Pco_2*.

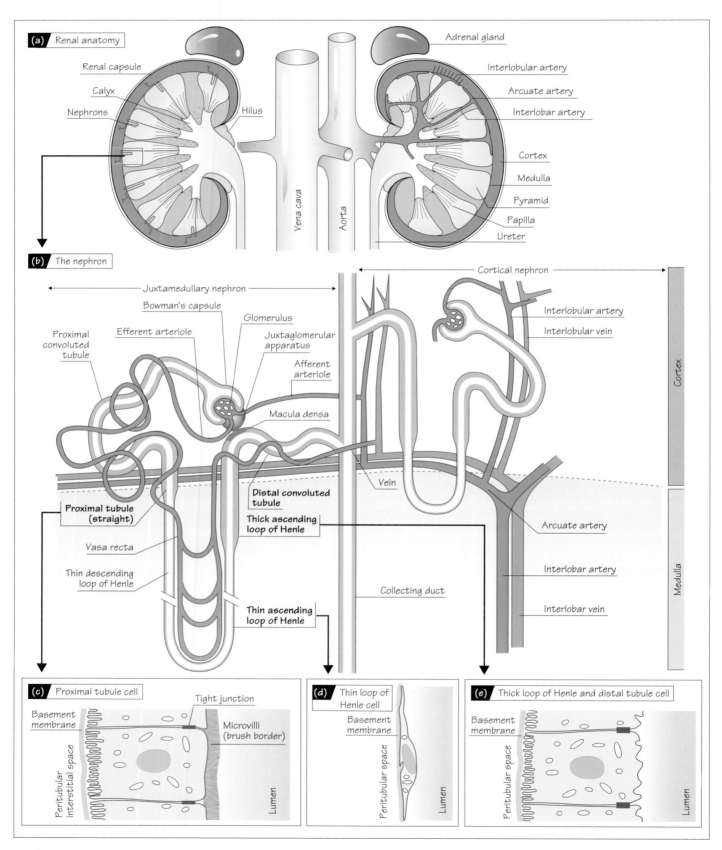

Physiology at a Glance, Third Edition. Jeremy P.T. Ward and Roger W.A. Linden.

 © 2013 John Wiley & Sons, Ltd. Published 2013 by John Wiley & Sons, Ltd.

The kidneys help to maintain the composition of extracellular body fluids, and regulate ions (e.g. Na^+, K^+, Ca^{2+}, Mg^{2+}), acid–base status and body water. They also have an endocrine function. Plasma is **filtered** by capillaries in the **glomerulus** (Chapter 32), and the composition of the filtrate is modified by **reabsorption** and **secretion** in the nephrons. The average urine output is ~1.5 L per day, although this can fall to <1 L per day and increase to nearly 20 L per day.

Gross structure

The kidneys are located on each side of the vertebral column, behind the peritoneum. The renal artery and vein, lymphatics and nerve enter the kidney via the **hilus**, from which the **renal pelvis**, which becomes the **ureter**, emerges (Fig. 31a). The kidney is surrounded by a fibrous **renal capsule**. Internally, the kidney has a dark outer **cortex** surrounding a lighter **medulla**, which contains triangular lobes or **pyramids**. The cortex contains the **glomerulus** and **proximal** and **distal tubules** of the **nephrons**, whilst the **loop of Henle** and **collecting ducts** descend into the medulla (Fig. 31b). Each kidney contains ~800 000 nephrons. The collecting ducts converge in the **papilla** at the apex of each pyramid, and empty into the **calyx** (plural: *calyces*) and thence renal pelvis. Urine is propelled through the ureter into the **bladder** by peristalsis.

The nephron

Each **nephron** begins with a capsule (**Bowman's capsule**) surrounding the glomerular capillaries, which collects filtrate (Fig. 32a), followed by the **proximal tubule**, **loop of Henle**, **distal tubule** and early **collecting duct** (Fig. 31b). There are two types of nephron – those with glomeruli in the outer 70% of the cortex and short loops of Henle (**cortical nephrons**: ~85%), and those with glomeruli close to the cortex–medulla boundary and long loops of Henle (**juxtamedullary nephrons**: ~15%).

The **glomerulus** produces **ultrafiltrate** from plasma (Chapter 32).

The **proximal tubule** is convoluted when it leaves the Bowman's capsule, but straightens before becoming the descending limb of the loop of Henle in the medulla. Its walls are formed from *columnar epithelial* cells with a **brush-border** of *microvilli* on the luminal surface that increases the surface area ~40-fold (Fig. 31c). **Tight junctions** close to the luminal side limit diffusion through gaps between cells. The basal or peritubular side of the cells shows considerable *interdigitation*, which increases the surface area. The term **lateral intercellular space** is often used to describe the space between the interdigitations and basement membrane, and between the bases of adjacent cells. The main function of the proximal tubule is **reabsorption** (Chapter 33).

The thin part of the **loop of Henle** (~20 μ m across) is formed from thin, flat (*squamous*) cells (Fig. 31d), with no microvilli. The **thick ascending loop of Henle** has columnar epithelial cells similar to the proximal tubule, but with few microvilli (Fig. 31e). At the point at which the loop associates with the **juxtaglomerular apparatus** (Chapter 35), after re-entering the cortex, the wall is formed from modified **macula densa** cells (Fig. 31b). The loop of Henle is important for the production of concentrated urine.

The **distal tubule** is functionally similar to the **cortical collecting duct**. Both contain cells similar to those in the thick ascending loop of Henle (Fig. 31e). In the collecting duct, these **principal cells** are interspersed with **intercalated cells** of different morphology and function; these play a role in acid–base balance (Chapter 36). The collecting duct plays an important role in water homeostasis (Chapter 35).

Renal circulation

The kidneys receive ~20% of cardiac output. The renal artery enters via the hilus and divides into **interlobar arteries** running between the pyramids to the cortex–medulla boundary, where they split into **arcuate arteries**. **Interlobular arteries** ascend into the cortex, and feed the **afferent arterioles** of the glomerulus (Fig. 31a,b). The capillaries of the glomerulus are the site of **filtration**, and drain into the **efferent arteriole** (*not* vein). Afferent and efferent arterioles provide the major resistance to renal blood flow. Efferent arterioles branch into a network of capillaries in the cortex around the proximal and distal tubules (**peritubular capillaries**). Capillaries close to the cortex–medulla boundary loop into the medulla to form the **vasa recta** surrounding the loop of Henle; this provides the only blood supply to the medulla. All capillaries drain into the renal veins. Ninety per cent of the blood entering the kidney supplies the cortex, giving a high blood flow (~500 mL/min/100 g) and a low arteriovenous O_2 difference (~2%). Medullary blood flow is less (20–100 mL/min/100 g).

Regulation of renal blood flow. Differential constriction of afferent and efferent arterioles strongly affects filtration (see above; Chapter 32). The kidneys exhibit a high degree of **autoregulation** (Fig. 32e), both by the **myogenic** response (Chapter 24) and via the macula densa, which detects high filtration rates and releases adenosine, which constricts afferent arterioles, so reducing filtration. Noradrenaline (norepinephrine) from renal sympathetic nerves constricts both afferent and efferent arterioles, and increases renin and thus the production of angiotensin II (a potent vasoconstrictor) (Chapter 35). Many peripheral vasoconstrictors (e.g. endothelin, angiotensin II) cause the release of vasodilating prostaglandins in the kidney, so protecting renal blood flow.

Hormones and the kidney

Renal function is affected by a variety of hormones that modulate the regulation of ions and water (e.g. **antidiuretic hormone, aldosterone**). **Renin** is produced by the juxtaglomerular apparatus and promotes the formation of angiotensin (Chapter 35). **Erythropoietin** is synthesized by interstitial cells in the cortex, and stimulates red cell production (Chapter 8). **Vitamin D** is metabolized in the kidney to its active form (**1,25-dihydroxycholecalciferol**), which is involved in Ca^{2+} and phosphate regulation (Chapters 34 and 48). Various **prostaglandins** are also produced in the kidney, and affect renal blood flow.

Micturition

The constriction of smooth muscle in the bladder wall (**detrusor muscle**) expels urine through the **urethra** (**micturition**, urination). Micturition is initiated by a spinal reflex when urine pressure reaches a critical level, but is strongly controlled by higher (voluntary) centres. The neck of the bladder forms the **internal urethral sphincter**; the **external sphincter** is formed from voluntary skeletal muscle around more distal regions of the urethra.

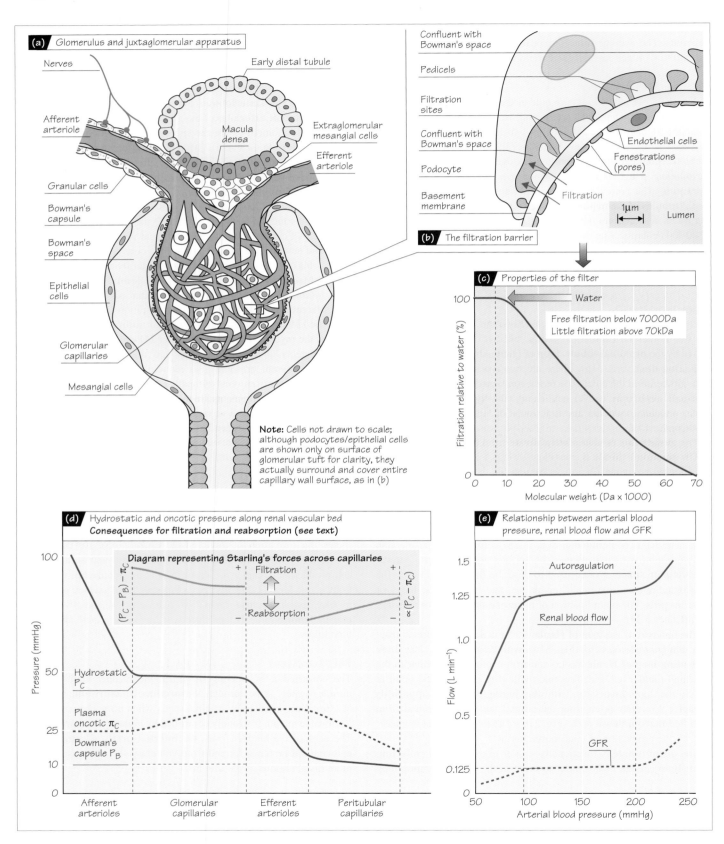

Physiology at a Glance, Third Edition. Jeremy P.T. Ward and Roger W.A. Linden.

 © 2013 John Wiley & Sons, Ltd. Published 2013 by John Wiley & Sons, Ltd.

The structure of the **glomerulus** is shown in Figure 32a. The walls of the afferent arteriole are associated with *granular cells* that produce **renin** (Chapter 35); there are numerous sympathetic nerve endings. The tuft of glomerular capillaries is surrounded by **Bowman's capsule**, the inner surface of which and the capillaries are covered by specialized epithelial cells (**podocytes**; see below). The glomerulus is interspersed with **mesangial** cells which are **phagocytic** (engulf large molecules) and **contractile**; contraction may limit the filtration area and alter filtration. Mesangial cells are also found between the capsule and **macula densa** (*extraglomerular mesangial cells*; Fig. 32a).

Glomerular filtration

Plasma is filtered in the glomerulus by **ultrafiltration** (i.e. works at the molecular level), and filtrate passes into the proximal tubule. The **glomerular filtration rate** (GFR) is ~125 mL/min in humans. The renal *plasma* flow is ~600 mL/min, so that the proportion of plasma that filters into the nephron (**filtration fraction**) is ~20%. Fluid and solutes have to pass three filtration barriers (Fig. 32b):

1 The **glomerular capillary endothelium**, which is approximately 50 times more permeable than in most tissues because it is **fenestrated** with small (70 nm) pores (Chapter 23).

2 A specialized capillary **basement membrane** containing negatively charged glycoproteins, which is thought to be the main site of ultrafiltration.

3 Modified epithelial cells (**podocytes**) with long extensions (*primary processes*) that engulf the capillaries and have numerous foot-like processes (**pedicels**) directly contacting the basement membrane. The regular gaps between pedicles are called **filtration slits**, and restrict large molecules. Podocytes maintain the basement membrane and, like mesangial cells, may be phagocytic and partially contractile.

The permeability of the filtration barrier is dependent on the molecular size. Substances with molecular weights of <7000 Da pass freely, but larger molecules are increasingly restricted up to 70000 Da, above which filtration is insignificant (Fig. 32c). Negatively charged molecules are further restricted as they are repelled by negative charges in the basement membrane. Thus, albumin (~69000 Da), which is also negatively charged, is filtered in minute quantities, whereas small molecules such as ions, glucose, amino acids and urea pass the filter without hindrance. This means that the glomerular filtrate is almost protein free, but otherwise has an identical composition to plasma.

Factors determining the glomerular filtration rate

GFR is dependent on the difference between the **hydrostatic** and **oncotic** (colloidal osmotic, due to proteins) pressures in the glomerular capillaries and Bowman's capsule, as determined by **Starling's equation** (Chapter 23). The glomerular capillary pressure (P_c) is greater than that elsewhere (~48 mmHg) because of the unique arrangement of afferent and efferent arterioles, and low afferent but high efferent resistances. As the pressure in Bowman's capsule (P_B) is ~10 mmHg, the net hydrostatic force driving filtration is ($P_c - P_B$) or ~35 mmHg.

This is opposed by the oncotic pressure of capillary plasma (π_c; ~25 mmHg); the filtrate oncotic pressure is essentially zero (no protein). Thus, GFR $\propto (P_c - P_B) - \pi_c$ (Fig. 32d). It should be noted that, because the filtration fraction is appreciable (~20%) and proteins are not filtered, the plasma protein concentration and thus π_c will rise as blood traverses the glomerulus, reducing (*but not abolishing*) filtration. In peritubular capillaries, where the hydrostatic pressure is very low, this increase in π_c promotes reabsorption (Fig. 32d).

GFR is therefore strongly dependent on the relative resistance of afferent and efferent arterioles, which is influenced by sympathetic tone and other vasoactive agents. GFR is constant over a wide range of blood pressure (90–200 mmHg) because of the **autoregulation** of renal blood flow (Fig. 32e; Chapter 24). Renal disease, circulating and local vasoconstrictors, and sympathetic activation all reduce GFR, although angiotensin II preferentially constricts *efferent* arterioles, and thus increases GFR (Chapter 35).

Measurement of the glomerular filtration rate and the concept of clearance

If substance X is freely filtered and neither reabsorbed nor secreted in the nephron, the amount appearing in the urine per minute must equal the amount filtered per minute. Thus, if the plasma concentration of X is C_p and the urine concentration is C_u, and the volume of urine passed per minute is V, then $C_p \times$ GFR $= C_u \times V$, or **GFR $= (C_u \times V)/C_p$**.

Creatinine, which is steadily released from skeletal muscle, is often used for clinical measurements of GFR because it is freely filtered and not reabsorbed; there is a little secretion, but this introduces only a small error, except when plasma creatinine or GFR is abnormally low. More accurate measurements are made by infusing the polysaccharide **inulin**, which is neither reabsorbed nor secreted.

This is known as a **clearance method**. The term **clearance** can be confusing, as it does not refer to what actually happens but is merely a way of looking at how the kidney deals with a substance. It is defined as the volume of plasma that would need to be completely cleared of a substance per minute in order to produce the amount found in the urine, or: **clearance $= (C_u \times V)/C_p$** (i.e. the same equation as above). Thus, the **clearance of inulin is equal to GFR**. If a substance is reabsorbed in the nephron, its clearance will be less than the GFR and, if it is secreted, it will be greater than the GFR. Some substances that are normally completely reabsorbed have zero clearance until the reabsorption mechanism becomes saturated (e.g. glucose; Chapter 33).

The **renal plasma flow** (RPF) can be measured in a similar fashion by infusing ***para*-aminohippuric acid** (**PAH**) which at low concentrations is completely removed from renal blood by both filtration and secretion, so that none remains in the venous outflow. The amount appearing in the urine must therefore equal the amount entering the kidney, and thus the **clearance of PAH is equal to RPF**. The filtration fraction (GFR/RPF; see above) can therefore be estimated from inulin clearance/PAH clearance. The renal blood flow is equal to RPF/ (1 − haematocrit).

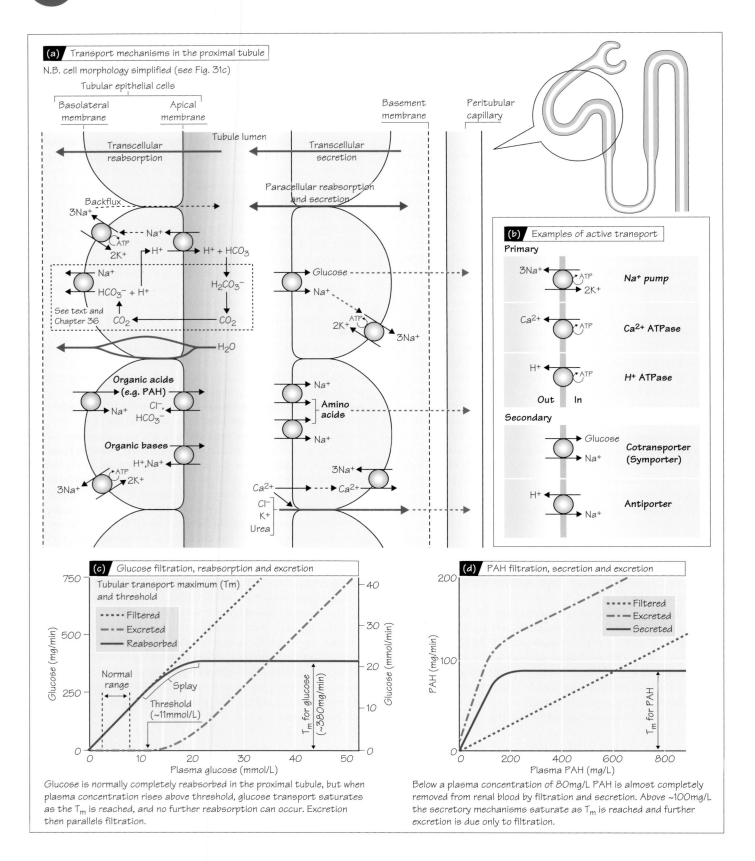

(c) Glucose filtration, reabsorption and excretion

Glucose is normally completely reabsorbed in the proximal tubule, but when plasma concentration rises above threshold, glucose transport saturates as the T_m is reached, and no further reabsorption can occur. Excretion then parallels filtration.

(d) PAH filtration, secretion and excretion

Below a plasma concentration of 80mg/L PAH is almost completely removed from renal blood by filtration and secretion. Above ~100mg/L the secretory mechanisms saturate as T_m is reached and further excretion is due only to filtration.

Physiology at a Glance, Third Edition. Jeremy P.T. Ward and Roger W.A. Linden.

76 © 2013 John Wiley & Sons, Ltd. Published 2013 by John Wiley & Sons, Ltd.

In a healthy adult, ~180 L of filtrate enters the proximal tubules daily. A significant component must be reabsorbed to prevent the loss of water and solutes. The filtrate is progressively modified as it passes through the nephron by the **reabsorption** of substances into the blood and **secretion** into the tubular fluid. The net reabsorption or secretion of any substance can be determined from its **clearance** (Chapter 32).

Tubular transport processes

Reabsorption and secretion involve the transport of substances across the tubular epithelium; this occurs either by diffusion through tight junctions and lateral intercellular spaces (**paracellular** pathway), driven by concentration, osmotic or electrical gradients, or by **active transport** through the epithelial cells themselves (**transcellular** pathways) (Fig. 33a). The latter usually involves an active process on either the apical or basolateral cell membrane, with passive diffusion across the opposite membrane driven by the concentration gradient so created. The movement of solutes between the peritubular space and capillaries is by **bulk flow and diffusion** (Chapter 11); the movement of water is influenced by **Starling's forces** (Chapter 23).

Active transport involves proteins called **transporters** that translocate substances across the cell membrane (Fig. 33b; Chapter 4). **Primary active transport** uses adenosine triphosphate (ATP) directly, e.g. the **Na^+–K^+ ATPase** (Na^+ pump). **Secondary active transport** uses the concentration gradient created by primary active transport as an energy source. This is most commonly the Na^+ gradient created by the *Na^+ pump*, and the latter therefore plays a critical role in renal reabsorption and secretion. **Symporters** (or *cotransporters*) transport substances in the same direction as (for example) Na^+, whereas **antiporters** transport in the opposite direction (Chapter 4; Fig. 33b).

The rate of *diffusion* across cell membranes is enhanced by **ion channels** and **uniporters** (transporters carrying only one substance), which effectively increase membrane permeability to specific substances; this is termed **facilitated diffusion**, and may be modulated by hormones or drugs.

Tubular transport maximum

There is a limit to the rate at which any transporter can operate, and so, for any substance, there is a maximum rate of reabsorption or secretion, called the **tubular transport maximum (T_m)**. For example, glucose is normally completely reabsorbed in the proximal tubule and none is excreted in the urine (see below). However, when the filtrate glucose concentration rises above the **renal threshold**, the transporters start to **saturate**, and glucose appears in the urine (Fig. 33c). Once T_m is reached, excretion increases linearly with filtration. The threshold concentration is somewhat lower than that required to reach T_m because of the variation in transport maxima between nephrons; this is called **splay**. Secretory mechanisms also exhibit T_m. For example, at low concentrations, *para*-aminohippuric acid (PAH) is almost completely removed from capillary blood by filtration and secretion (Chapter 32). At higher concentrations secretion becomes saturated, and further excretion is limited to the filtered load (Fig. 33d).

The proximal tubule (Fig. 33a)

Most glucose, amino acids, phosphate and bicarbonate is reabsorbed in the proximal tubule, together with 60–70% of Na^+, K^+, Ca^{2+}, urea and water. The secretion of H^+ and reabsorption of HCO_3^- are discussed in detail in Chapter 36.

Sodium. The concentration of Na^+ in the filtrate is ~140 mmol/L (= plasma Na^+ concentration), but is much lower in the cytosol of epithelial cells (~10–20 mmol/L), which is also negatively charged. The electrochemical gradient therefore favours the movement of Na^+ from the filtrate into the cells, providing the driving force for the secondary transport of other substances. About 80% of Na^+ entering proximal tubular cells exchanges for H^+ (Na^+–H^+ antiporter). The secretion of H^+ in the proximal tubule plays a critical role in HCO_3^- reabsorption (Fig. 33a; Chapter 36). Na^+ is removed from tubular cells by Na^+ pumps primarily on the basolateral membrane, thus transporting Na^+ into the interstitial fluid. However, only ~20% of transported Na^+ diffuses into the capillaries, as there is significant backflux into the tubule via paracellular pathways.

Water. Water is not actively reabsorbed. As Na^+ and HCO_3^- are transported from the tubule into the peritubular interstitial fluid, the **osmolality** of the latter increases, whilst that of the tubular fluid decreases. This osmotic pressure difference causes the reabsorption of water via both transcellular and paracellular pathways.

The reabsorption of water increases tubular concentrations of Cl^-, K^+, Ca^{2+} and **urea**, which therefore diffuse down their concentration gradients into the peritubular space, largely via paracellular pathways, although the route for Ca^{2+} may be transcellular. The final two-thirds of the proximal tubule has increased permeability to Cl^-, facilitating Cl^- reabsorption. This makes the lumen more positive, enhancing the reabsorption of cations. As the reabsorption of Na^+, Cl^-, K^+, Ca^{2+} and urea in the proximal tubule is closely coupled to the reabsorption of water, their concentrations (and the total osmolality) are similar in the fluid leaving the proximal tubule to those in the filtrate and plasma, although their total quantity and the fluid volume are decreased by ~70%.

Glucose. Glucose is reabsorbed by **cotransport** with Na^+ across the apical membrane of epithelial cells, and then diffuses out of the cells into the peritubular interstitium. The T_m for glucose is ~380 mg/min (~21 mmol/min), and the renal threshold is ~11 mmol L^{-1}. The appearance of glucose in the urine reflects **hyperglycaemia** (high plasma glucose), a sign of **diabetes mellitus**.

Amino acids. Amino acids are reabsorbed by several Na^+-linked symporters, specific for acidic, basic and neutral amino acids.

Phosphate. Phosphate is cotransported with Na^+ across the epithelial apical membrane. Its T_m is close to the filtered load, and so an increase in plasma concentration leads to excretion. Phosphate reabsorption is decreased by **parathyroid hormone**.

Organic acids and bases. These include metabolites (e.g. bile salts, urate, oxalate) and drugs (e.g. PAH, penicillins, aspirin) and are secreted. Organic acids are transported from the peritubular fluid into tubular cells by cotransport with Na^+, and diffuse into the tubule in exchange for anions (e.g. Cl^-, HCO_3^-). Organic bases are actively extruded from the apical membrane in exchange for Na^+ or H^+.

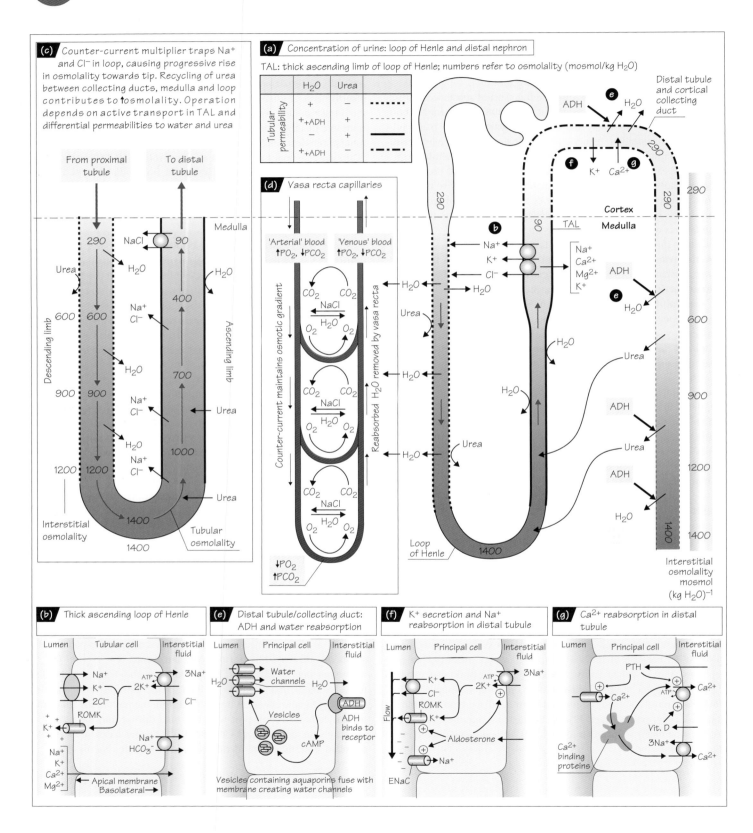

(c) Counter-current multiplier traps Na⁺ and Cl⁻ in loop, causing progressive rise in osmolality towards tip. Recycling of urea between collecting ducts, medulla and loop contributes to ↑osmolality. Operation depends on active transport in TAL and differential permeabilities to water and urea

(a) Concentration of urine: loop of Henle and distal nephron

TAL: thick ascending limb of loop of Henle; numbers refer to osmolality (mosmol/kg H_2O)

Tubular permeability	H_2O	Urea	
	+	−	·······
	+ +ADH	+	- - - - -
	−	+	────
	+ +ADH	−	—·—·—

(d) Vasa recta capillaries

(b) Thick ascending loop of Henle

(e) Distal tubule/collecting duct: ADH and water reabsorption

(f) K⁺ secretion and Na⁺ reabsorption in distal tubule

(g) Ca^{2+} reabsorption in distal tubule

Physiology at a Glance, Third Edition. Jeremy P.T. Ward and Roger W.A. Linden.

© 2013 John Wiley & Sons, Ltd. Published 2013 by John Wiley & Sons, Ltd.

The loop of Henle and distal nephron allow urine to be **concentrated** through the creation of a **high osmolality** in the medulla, which drives the reabsorption of water from the **collecting ducts**. The distal nephron also regulates K^+ and Ca^{2+} excretion and acid–base status (Chapter 36).

The loop of Henle

Fluid entering the descending limb of the loop of Henle is isotonic with plasma (~290 mosmol/kg H_2O). The generation of high osmolality in the medulla depends on the **differential permeabilities** to water and solutes in different regions, the **active transport** of ions in the thick ascending limb and the **counter-current multiplier**. The **thin descending limb** is permeable to water but impermeable to urea, whereas the **ascending limb** is impermeable to water but permeable to urea (Fig. 34a); it is also very highly permeable to Na^+ and Cl^-. The **thick ascending limb** actively reabsorbs Na^+ and Cl^- from the tubular fluid by means of apical Na^+–K^+–$2Cl^-$ **cotransporters**; Na^+ is primarily transported across the basolateral membrane by Na^+ pumps (some by Na^+–HCO_3^- cotransport), and Cl^- by diffusion (Fig. 34b and ❶). K^+ leaks back into the lumen via apical **ROMK** K^+ channels, creating a positive charge that drives the reabsorption of cations (Na^+, K^+, Ca^{2+}, Mg^{2+}) through paracellular pathways. As the thick ascending limb is impermeable to water, the reabsorption of ions reduces the tubular fluid osmolality (to ~90 mosmol/kg H_2O) and increases the interstitial fluid osmolality, creating an osmotic difference of **~200 mosmol/kg H_2O**.

Counter-current multiplier (Fig. 34c). The increased interstitial osmolality causes water to diffuse out of the descending limb, and some Na^+ and Cl^- to diffuse in, concentrating the tubular fluid (Fig. 34c). As this concentrated fluid descends, it travels in the opposite direction to fluid returning from the still higher osmolality regions of the deep medulla. This **counter-current** arrangement creates an osmotic gradient, causing Na^+ and Cl^- to diffuse out of the ascending limb (diluting the ascending fluid), and water to diffuse out of the descending limb (further concentrating the descending fluid). This effect is potentiated by the fact that the ascending limb is impermeable to water, but highly permeable to Na^+ and Cl^-, and also by the recycling of **urea** between the collecting ducts and ascending limb, which makes an important contribution to urine concentration (see below). At the tip of the loop of Henle, the interstitial fluid can reach an osmolality of **~1400 mosmol/kg H_2O**, due in equal parts to NaCl and urea.

The blood supply to the medulla is prevented from dissipating the osmotic gradient between the cortex and medulla by the *counter-current exchanger* arrangement of the **vasa recta** capillaries (Fig. 34d). The vasa recta also removes water reabsorbed from the loop of Henle and medullary collecting ducts. It should be noted that O_2 and CO_2 are also conserved, so that, in the deep medulla, P_{O_2} is low and P_{CO_2} is high.

The distal tubule and collecting duct

Fluid entering the distal tubule is *hypotonic* (~90 mosmol/kg H_2O). More Na^+ is reabsorbed in principal cells via the Na^+ channel **ENaC**, which is inhibited by atrial natriuretic peptide (**ANP**); expression of ENaC and thus Na^+ reabsorption is increased by **aldosterone** (Chapter 35). The movement of Na^+ through ENaC is charge compensated by

the opposite movement of K^+ through ROMK (Fig. 34f and ❶). The distal tubule and cortical collecting duct are impermeable to urea. They are also impermeable to water, except in the presence of **antidiuretic hormone** (ADH, *vasopressin*) (Chapter 35), which causes water channels (**aquaporins**) to insert into the apical membrane (Fig. 34e and ❸). In the presence of ADH, water diffuses into the cortical interstitium, and the tubular fluid becomes concentrated, reaching a maximum osmolality of ~290 mosmol/kg H_2O (i.e. isotonic with plasma). However, the fluid differs from plasma as large quantities of Na^+, K^+, Cl^- and HCO_3^- have been reabsorbed, their place having being taken by **urea**. This is concentrated as water is reabsorbed, because the distal tubule and cortical collecting duct are impermeable to urea.

The **medullary collecting duct** also becomes permeable to water in the presence of ADH. Water is reabsorbed due to the high osmolality of the medullary interstitium (Fig. 34a). The final urine osmolality can therefore reach **1400 mosmol/kg H_2O** under conditions of maximum ADH stimulation; in the absence of ADH, urine is *dilute* (**~60 mosmol/kg H_2O**) (Chapter 35). Although only 15% of nephrons have loops of Henle that pass deep into the medulla, and so contribute to the high medullary osmolality (Chapter 31), the *collecting ducts of all nephrons pass through the medulla and therefore concentrate urine.*

Urea. The **medullary collecting duct** is relatively permeable to urea, which diffuses down its concentration gradient into the medulla and then into the ascending loop of Henle (Fig. 34a). Urea is therefore 'trapped' and partially recycled, so maintaining a high concentration and providing ~50% of the osmolality in the medulla (see above). ADH increases the permeability of the medullary collecting duct to urea and hence its reabsorption by activating epithelial **uniporters** (*facilitated diffusion*); this further increases the medullary osmolality and allows the production of more concentrated urine.

Potassium. Potassium has largely been reabsorbed by the time the distal tubule is reached, and so excretion is regulated by secretion in the late distal tubule. K^+ is actively transported into principal cells by basolateral Na^+ pumps, and passively secreted via **ROMK channels** and K^+–Cl^- **cotransport**; the former is promoted by the negative luminal charge caused by reabsorption of Na^+ through ENaC (Fig. 34f and ❶). Secretion is therefore driven by the concentration gradient between the cytosol and tubular fluid. However, secreted K^+ will reduce the gradient unless it is washed away, and so **K^+ excretion is increased as tubular flow increases**. Diuretics therefore often lead to K^+ loss (Chapter 36). K^+ secretion is increased by **aldosterone**, which enhances Na^+ pump activity and apical membrane K^+ permeability (Chapter 35). Perturbations of K^+ homeostasis are often associated with acid–base disorders (Chapter 36).

Calcium. Calcium reabsorption in the distal tubule is regulated by **parathyroid hormone** (**PTH**) and **1,25-dihydroxycholecalciferol** (active form of **vitamin D**). PTH activates Ca^{2+} entry channels in the epithelial apical membrane, and a basolateral Ca^{2+} ATPase that is also activated by 1,25-dihydroxycholecalciferol. Ca^{2+} removal is assisted by an Na^+–Ca^{2+} antiporter. Ca^{2+}-binding proteins prevent cytosolic free Ca^{2+} from rising detrimentally (Fig. 34g and ❾). PTH also inhibits phosphate reabsorption (Chapter 33). Ca^{2+} regulation is discussed in Chapter 48.

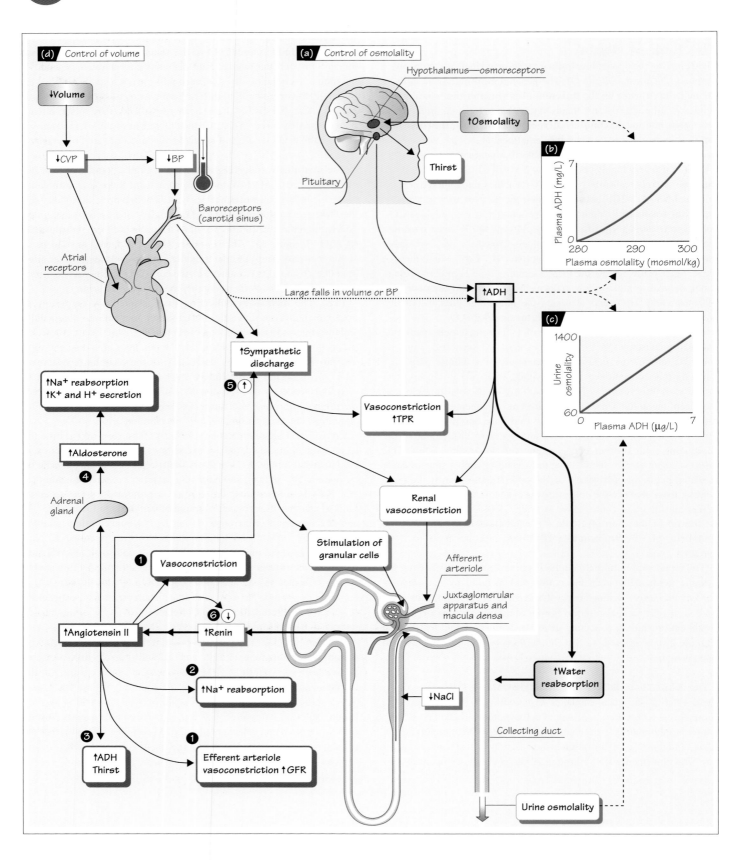

Physiology at a Glance, Third Edition. Jeremy P.T. Ward and Roger W.A. Linden.

 © 2013 John Wiley & Sons, Ltd. Published 2013 by John Wiley & Sons, Ltd.

Control of plasma osmolality (Fig. 35a)

Extracellular fluid osmolality must be closely regulated, as alterations cause the swelling or shrinking of all cells, and can lead to cell death. The control of osmolality takes precedence over the control of body fluid volume.

Plasma osmolality is increased in water deficiency and decreased by the ingestion of water. **Osmoreceptors** in the **anterior hypothalamus** are sensitive to changes as small as 1% of plasma osmolality, and regulate **antidiuretic hormone** (ADH), also known as **vasopressin**. A rise in osmolality increases ADH release and stimulates **thirst** and water reabsorption; a fall has the opposite effect. ADH is a peptide of nine amino acids formed from a large precursor synthesized in the **hypothalamus** (Chapter 44). ADH is transported from there to the **posterior pituitary** (*neurohypophysis*) within nerve fibres (*hypothalamohypophyseal tract*), where it is stored in **secretory granules**. Action potentials from osmoreceptors cause these to release ADH. ADH binds to V_2 receptors on renal principal cells and increases cyclic adenosine monophosphate (cAMP), causing the incorporation of water channels (*aquaporins*) into the apical membrane (Chapter 34). ADH also causes **vasoconstriction** (including renal) via V_1 receptors.

The relationship between plasma osmolality and ADH release is steep (Fig. 35b), as is the relationship between plasma ADH and urine osmolality (Fig. 35c). Normal urine production is ~60 mL/h (urine osmolality, ~300–800 mosmol/kg H_2O). Maximum ADH reduces the urine volume to a **minimum** of ~400 mL per day (**maximum urine osmolality**, ~1400 mosmol/kg H_2O; this cannot be greater than that in the deep medulla, Chapter 34). In the absence of ADH, the urine volume can reach ~25 L per day with a **minimum urine osmolality** of ~60 mosmol/kg H_2O (Chapter 34). ADH is rapidly removed from plasma, falling by ~50% in ~10 min, mainly due to metabolism in the liver and kidneys.

Diabetes insipidus is the production of copious amounts of **hypotonic** (dilute) urine due to defective ADH-dependent water reabsorption. This may be due to a congenital defect in ADH production (*central diabetes insipidus, CDI*), or to a failure to respond to ADH (*nephrogenic diabetes insipidus, NDI*) due to defective ADH receptors or aquaporins.

Control of body fluid volume (Fig. 35d)

As plasma osmolality is strongly regulated by the osmoreceptors and ADH, changes in the major osmotic component of extracellular fluid, i.e. Na^+, will result in changes in extracellular volume. The control of body Na^+ content by the kidney is therefore the main regulator of body fluid volume. Atrial and other low-pressure (cardiopulmonary) stretch receptors (Fig. 35d) detect a fall in central venous pressure (**CVP**), which reflects the blood volume. A fall in volume sufficient to reduce blood pressure activates the **baroreceptor reflex** (Chapter 22). In both cases, increased sympathetic discharge causes peripheral vasoconstriction (increasing total peripheral resistance; TPR),, including vasoconstriction of the **renal afferent arterioles**, stimulation of **ADH release** and water reabsorption (see above), and **release of renin** (see below) from **granular cells** in the juxtaglomerular apparatus (Chapter 31). **Decreased pressure** in the renal afferent arterioles also stimulates renin release, as does reduced NaCl delivery to the **macula densa** in the juxtaglomerular apparatus (Chapter 31) and a reduced glomerular filtration rate (GFR). In extremis, *large* falls in blood volume or pressure will promote ADH release and water retention at the expense of a decreased plasma osmolality. This only occurs where the alternative is circulatory failure, and is not sustainable.

Renin, angiotensin and aldosterone

Renin cleaves plasma angiotensinogen into angiotensin I, which is converted by **angiotensin-converting enzyme** (ACE) on endothelial cells (primarily in the lung) into **angiotensin II**. Angiotensin II is the primary hormone for Na^+ homeostasis, and has several important functions (Fig. 35d). It is a potent **vasoconstrictor** throughout the vasculature ❶, although in the kidney it preferentially constricts efferent arterioles, thereby increasing GFR (Chapter 32) and protecting GFR from a fall in perfusion pressure. It directly increases **Na^+ reabsorption** in the proximal tubule ❷ by stimulating Na^+–H^+ antiporters; (Chapter 33). It stimulates the hypothalamus to increase **ADH secretion** and also causes **thirst** ❸. It stimulates the production of **aldosterone** by the adrenal cortex ❹. Angiotensin II also tends to potentiate sympathetic activity ❺ (*positive feedback*) and inhibit renin production by granular cells ❻ (*negative feedback*). **ACE inhibitors** are important for the treatment of heart failure, when the response to reduced blood pressure leads to detrimental fluid retention and oedema (Chapter 23).

Aldosterone is required for normal Na^+ reabsorption and K^+ secretion. It increases the synthesis of transport mechanisms in the distal nephron, including the Na^+ pump, Na^+–H^+ symporter and K^+ and Na^+ channels in principal cells, and H^+ ATPase in intercalated cells. Na^+ reabsorption and K^+ and H^+ secretion are thereby enhanced (Chapters 34 and 36). As aldosterone acts via **protein synthesis**, it takes hours to have any effect. The production of aldosterone by the adrenal cortex is directly sensitive to small changes in **plasma [K^+]**, suggesting a primary role for K^+ homeostasis.

Atrial natriuretic peptide (ANP; *atrial natriuretic factor*) is released from atrial muscle cells in response to stretch caused by increased blood volume (Chapter 22). ANP inhibits ENaC in principal cells of the distal nephron (Chapter 34), suppresses the production of renin, aldosterone and ADH, and causes renal vasodilatation. The net result is increased excretion of water and Na^+.

Diuretics

Osmotic diuretics (e.g. mannitol) cannot be reabsorbed effectively and, consequently, their concentration in tubular fluid increases as water is reabsorbed, limiting further water reabsorption. In **diabetes mellitus**, high plasma glucose saturates glucose reabsorption (Chapter 33), resulting in copious amounts of isotonic urine (i.e. same osmolality as plasma) containing glucose. **Diuretic drugs** generally inhibit tubular transport mechanisms. The most potent are **loop diuretics** (e.g. furosemide), which inhibit Na^+–K^+–$2Cl^-$ symporters in the thick ascending loop of Henle, thus preventing the development of high osmolality in the medulla and inhibiting water reabsorption (Chapter 34). The increased flow (and thus increased K^+ secretion), coupled with reduced K^+ reabsorption, enhances K^+ excretion and can cause **hypokalaemia** (low plasma [K^+]). **Aldosterone antagonists** (e.g. spironolactone) and **Na^+ channel blockers** (e.g. amiloride) reduce Na^+ entry in the distal nephron and inhibit K^+ and H^+ secretion; they are weak diuretics, but **K^+ sparing**, and are often given with loop diuretics to reduce K^+ loss. **Alcohol** inhibits ADH release, and so promotes diuresis.

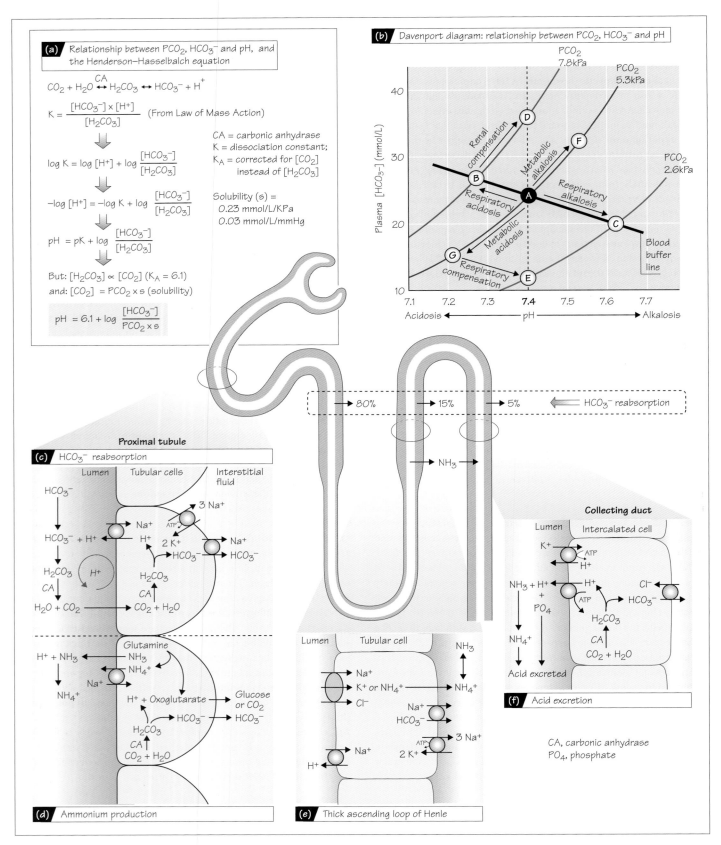

(a) Relationship between PCO_2, HCO_3^- and pH, and the Henderson–Hasselbalch equation

$$CO_2 + H_2O \overset{CA}{\longleftrightarrow} H_2CO_3 \longleftrightarrow HCO_3^- + H^+$$

$$K = \frac{[HCO_3^-] \times [H^+]}{[H_2CO_3]} \quad \text{(From Law of Mass Action)}$$

$$\log K = \log [H^+] + \log \frac{[HCO_3^-]}{[H_2CO_3]}$$

$$-\log [H^+] = -\log K + \log \frac{[HCO_3^-]}{[H_2CO_3]}$$

$$pH = pK + \log \frac{[HCO_3^-]}{[H_2CO_3]}$$

But: $[H_2CO_3] \propto [CO_2]$ ($K_A = 6.1$)
and: $[CO_2] = PCO_2 \times s$ (solubility)

$$pH = 6.1 + \log \frac{[HCO_3^-]}{PCO_2 \times s}$$

CA = carbonic anhydrase
K = dissociation constant;
K_A = corrected for $[CO_2]$ instead of $[H_2CO_3]$

Solubility (s) =
0.23 mmol/L/KPa
0.03 mmol/L/mmHg

(b) Davenport diagram: relationship between PCO_2, HCO_3^- and pH

HCO_3^- reabsorption — 80% — 15% — 5%

Proximal tubule

(c) HCO_3^- reabsorption

(d) Ammonium production

(e) Thick ascending loop of Henle

Collecting duct

(f) Acid excretion

CA, carbonic anhydrase
PO_4, phosphate

Physiology at a Glance, Third Edition. Jeremy P.T. Ward and Roger W.A. Linden.
© 2013 John Wiley & Sons, Ltd. Published 2013 by John Wiley & Sons, Ltd.

The pH of arterial blood is 7.35–7.45 ($[H^+]$ = 45–35 nmol/L). Metabolism produces ~60 mmol H^+ per day, most of which is excreted through the lungs as CO_2, formed by the reaction of H^+ with HCO_3^- (bicarbonate) (Fig. 36a). The kidneys conserve and replace HCO_3^- lost in this way, and fine tune H^+ excretion. Physiological **buffers** maintain a low *free* $[H^+]$ and prevent large swings in pH.

Buffers

Buffers are **weak acids** (HA) or **bases** (A^-) that can donate or accept H^+ ions. The ratio between buffer pairs (e.g. carbonic acid, H_2CO_3, and bicarbonate, HCO_3^-) is determined by $[H^+]$ and the **dissociation constant** (K) for that buffer pair: K = $([H^+][A^-])/[HA]$, or pH = pK + $\log([A^-]/[HA])$ (the **Henderson–Hasselbalch equation**). Thus, an increase in $[A^-]$ or a decrease in [HA] will increase pH (more alkaline), and a decrease in pH will decrease the ratio $[A^-]/[HA]$. Buffers work best when the pH is close to their **pK value**, the pH at which the ratio $[A^-]/[HA]$ is unity. Bicarbonate and carbonic acid (formed by the combination of CO_2 with water, greatly potentiated by **carbonic anhydrase**, CA; Fig. 36a) are the most important buffer pair in the body, although haemoglobin provides ~20% of buffering in the blood; phosphate and proteins provide intracellular buffering. Buffers in urine, largely phosphate, allow the excretion of large quantities of H^+.

Although the HCO_3^- system has a pK value of 6.1, and is theoretically a poor buffer at pH 7.4, it is physiologically effective because CO_2 (and therefore H_2CO_3) and HCO_3^- are precisely controlled by the lungs (Chapter 29) and kidney, respectively. These fix the HCO_3^-/H_2CO_3 ratio and therefore the pH, and the latter determines the ratio of all other buffer pairs. The relationship between pH, P_{CO_2} and $[HCO_3^-]$ is described in Figure 36a,b. The line BAC is the **buffer line** for whole blood; changes in P_{CO_2} alter HCO_3^- and pH along this line. Point A denotes normal conditions (pH 7.4, $[HCO_3^-]$ = 24 mM, P_{CO_2} = 5.3 kPa).

Proximal renal tubule

Bicarbonate is freely filtered, and so filtrate $[HCO_3^-]$ is ~24 mmol/L (as in plasma). Less than 0.1% of filtered HCO_3^- is normally excreted in the urine, ~80% being reabsorbed in the proximal tubule. HCO_3^- is not transported directly. Filtered HCO_3^- associates with H^+ secreted by epithelial **Na^+–H^+ antiporters** to form H_2CO_3, which rapidly dissociates to CO_2 and H_2O in the presence of **carbonic anhydrase**. CO_2 and H_2O diffuse into the tubular cells, where they recombine into H_2CO_3, which dissociates to H^+ and HCO_3^-. HCO_3^- is transported into the interstitium largely by **Na^+–HCO_3^- symporters** (Fig. 36c). For each H^+ secreted into the lumen, one HCO_3^- and one Na^+ enter the plasma. H^+ is recycled, so that there is little net H^+ secretion at this stage. A further 10–15% of HCO_3^- is similarly reabsorbed in the thick ascending loop of Henle. In total, about 4000–5000 mmol of HCO_3^- is reabsorbed per day.

Ammonia is produced in tubular cells by the metabolism of glutamine, which leads to the generation of HCO_3^- and glucose or CO_2. NH_3 diffuses into the tubular fluid, or as NH_4^+ is transported by the Na^+–H^+ antiporter. In the tubular fluid, NH_3 gains H^+ to form NH_4^+, which cannot diffuse through membranes (Fig. 36d). About 50% of NH_4^+ secreted by the proximal tubule is reabsorbed in the thick ascending loop of Henle, where it substitutes for K^+ in the Na^+–K^+–$2Cl^-$ symporter (Chapter 34), and passes into the medullary interstitium (Fig. 36e). Here, NH_4^+ dissociates into NH_3 and H^+, and NH_3 re-enters the collecting duct by diffusion. The secretion of H^+ in the collecting duct (see below) leads to conversion back to NH_4^+, which is trapped in the lumen and excreted.

Distal renal tubule

The secretion of H^+ in the distal tubule promotes the reabsorption of any remaining HCO_3^-. The combination of H^+ with NH_3 (see above) and phosphate prevents H^+ recycling and allows acid excretion. In the early distal nephron, H^+ secretion is predominantly by Na^+–H^+ exchange, but more distally secretion is via **H^+ ATPase** and **H^+–K^+ ATPase** in **intercalated cells**, which contain plentiful carbonic anhydrase. As secreted H^+ is derived from CO_2, HCO_3^- is formed and returns to the blood (Fig. 36f).

In summary, in the proximal nephron, H^+ secretion promotes HCO_3^- reabsorption. In the distal nephron, secretion leads to the combination of H^+ with urinary buffers (phosphate, NH_3), and thus the generation of HCO_3^- and acid excretion. As a result of this, tubular fluid becomes more acid as it moves through the nephron. H^+ secretion is proportional to intracellular $[H^+]$, which is itself related to extracellular pH. A fall in blood pH will therefore stimulate renal H^+ secretion.

Acid–base regulation and compensation

Respiratory acidosis and **alkalosis** refer to alterations in pH caused by changes in P_{CO_2} (i.e. ventilation). **Metabolic acidosis** and **alkalosis** refer to changes not related to P_{CO_2} (i.e. increased acid production, diet, renal disease). Thus, hypoventilation increases P_{CO_2} and causes respiratory acidosis, denoted by the move from A to B in Fig. 36b. A *sustained* respiratory acidosis (e.g. *respiratory failure*) can be **compensated** by increased renal excretion of H^+ and reabsorption of HCO_3^-. The $[HCO_3^-]$/P_{CO_2} ratio is thus restored, and the pH returns towards normal. This **renal compensation** is denoted by the arrow B–D in Figure 36b. Similarly, **metabolic acidosis** (G) may be compensated by increased ventilation and reduced P_{CO_2} (G–E) (**respiratory compensation**), initiated by the detection of acid pH by the chemoreceptors (Chapter 29). Renal mechanisms are slow because their capacity for handling H^+ and HCO_3^- is smaller than that of the lungs for handling CO_2.

K^+ homeostasis and acid–base status

Hypokalaemia (low plasma $[K^+]$) is associated with metabolic alkalosis, due to stimulation of ammonia production, Na^+–H^+ exchange and H^+–K^+ ATPase, all of which enhance H^+ secretion. This is potentiated by aldosterone (Chapter 35). Hyperkalaemia has the opposite effect, and inhibits NH_4^+ reabsorption by competition at the Na^+–K^+–$2Cl^-$ symporter. Changes in acid–base status can affect K^+ homeostasis for similar reasons.

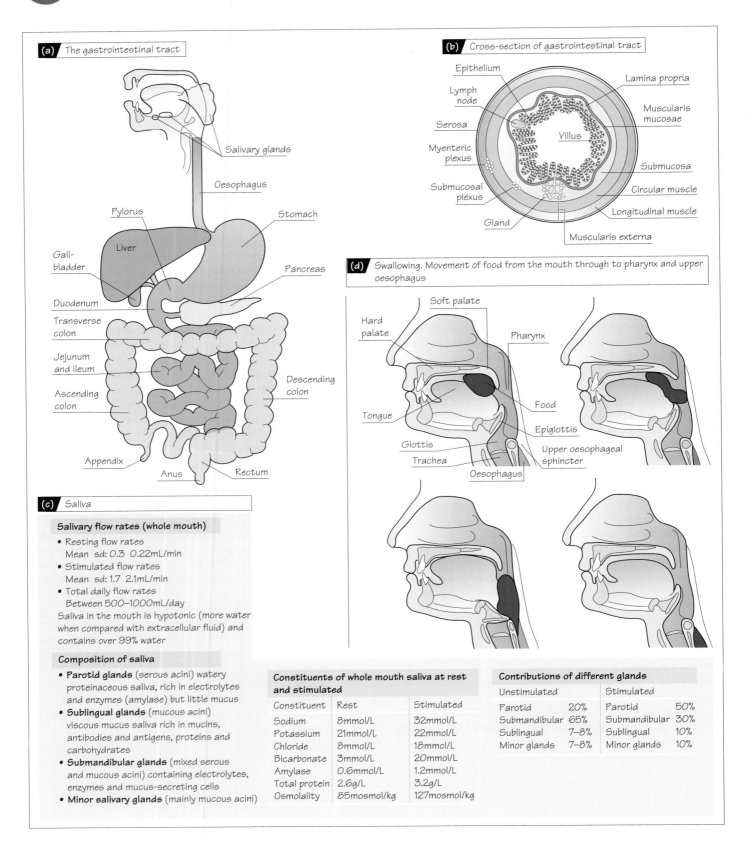

(a) The gastrointestinal tract

Salivary glands
Oesophagus
Pylorus
Stomach
Liver
Gall-bladder
Pancreas
Duodenum
Transverse colon
Jejunum and ileum
Ascending colon
Descending colon
Appendix
Anus
Rectum

(b) Cross-section of gastrointestinal tract

Epithelium
Lymph node
Serosa
Myenteric plexus
Submucosal plexus
Gland
Villus
Lamina propria
Muscularis mucosae
Submucosa
Circular muscle
Longitudinal muscle
Muscularis externa

(d) Swallowing. Movement of food from the mouth through to pharynx and upper oesophagus

Hard palate
Soft palate
Pharynx
Tongue
Food
Epiglottis
Glottis
Trachea
Upper oesophageal sphincter
Oesophagus

(c) Saliva

Salivary flow rates (whole mouth)
- Resting flow rates
  Mean sd: 0.3 0.22mL/min
- Stimulated flow rates
  Mean sd: 1.7 2.1mL/min
- Total daily flow rates
  Between 500–1000mL/day

Saliva in the mouth is hypotonic (more water when compared with extracellular fluid) and contains over 99% water

Composition of saliva
- **Parotid glands** (serous acini) watery proteinaceous saliva, rich in electrolytes and enzymes (amylase) but little mucus
- **Sublingual glands** (mucous acini) viscous mucus saliva rich in mucins, antibodies and antigens, proteins and carbohydrates
- **Submandibular glands** (mixed serous and mucous acini) containing electrolytes, enzymes and mucus-secreting cells
- **Minor salivary glands** (mainly mucous acini)

Constituents of whole mouth saliva at rest and stimulated

Constituent	Rest	Stimulated
Sodium	8mmol/L	32mmol/L
Potassium	21mmol/L	22mmol/L
Chloride	8mmol/L	18mmol/L
Bicarbonate	3mmol/L	20mmol/L
Amylase	0.6mmol/L	1.2mmol/L
Total protein	2.6g/L	3.2g/L
Osmolality	85mosmol/kg	127mosmol/kg

Contributions of different glands

Unstimulated		Stimulated	
Parotid	20%	Parotid	50%
Submandibular	65%	Submandibular	30%
Sublingual	7–8%	Sublingual	10%
Minor glands	7–8%	Minor glands	10%

Physiology at a Glance, Third Edition. Jeremy P.T. Ward and Roger W.A. Linden.
 © 2013 John Wiley & Sons, Ltd. Published 2013 by John Wiley & Sons, Ltd.

The **gastrointestinal (GI) tract** is responsible for the breakdown of food into its component parts so that they can be absorbed into the body. It is made up of the **mouth, oesophagus, stomach** and **small and large intestines**. The **salivary glands, liver, gallbladder** and **pancreas** are organs distinct from the GI tract, but all secrete juices into the tract and aid the digestion and absorption of the food (Fig. 37a).

Structure

Different regions of the tract are concerned with **motility** (transport), **storage, digestion, absorption** and **elimination of waste**, and these functions of the GI tract are controlled by **neuronal, hormonal** and **local regulatory mechanisms**.

The walls of the GI tract have a general structure that is similar along most of its length, although this is modified as function varies. This basic structure is shown in Figure 37b. It comprises the **mucosal layer**, made up of epithelial cells (which can be involved in either the process of secretion or absorption depending on their location in the GI tract), and the **lamina propria**, consisting of loose connective tissue, collagen and elastin, blood vessels and lymph tissue, and a thin layer of smooth muscle called the **muscularis mucosa** which, when contracting, produces folds and ridges in the mucosa. The **submucosal layer** comprises a second layer of connective tissue, but also contains larger blood and lymphatic vessels and a network of nerve cells called the **submucosal plexus (Meissner's plexus)**. This is a dense plexus of nerves innervated by the autonomic part of the nervous system which can function as an independent nervous system – the **enteric nervous system**. Below the submucosa is the **muscularis externa**. This comprises a thick **circular layer** of smooth muscle around the GI tract which, when it contracts, produces a constriction of the lumen. Below this layer of muscle is another thinner layer of muscle arranged in a **longitudinal** manner which, when it contracts, results in shortening of the tract. Between these two layers of muscle is a second nerve plexus, called the **myenteric plexus (Auerbach's plexus)**, which is also part of the enteric nervous system. The outermost layer of the GI tract is the **serosa**, another connective tissue layer covered with squamous mesothelial cells.

Saliva and mastication

The GI tract starts in the mouth, where food is initially **chewed (masticated)** and mixed with salivary secretions. **Mastication** is the process of systematic mechanical breakdown of food in the mouth. The amount of mastication necessary in order to swallow the food depends on the nature of the ingested food: solid foods are subjected to vigorous chewing, whereas softer foods and liquids require little or no chewing and are transported almost directly into the oesophagus by swallowing. Mastication is necessary for some foods, such as red meats, chicken and vegetables, to be fully absorbed by the rest of the GI tract. However, fish, eggs, rice, bread and cheese do not require chewing for complete absorption in the tract.

Mastication involves the coordinated activity of the **teeth, jaw muscles, temporomandibular joint, tongue** and other structures, such as the **lips, palate** and **salivary glands**. The forces developed between the teeth during mastication have been measured to be about 150–200 N; however, the maximum biting force developed between the molar teeth is almost 10 times this value.

During mastication three pairs of glands, the **parotid, submandibular** and **sublingual**, secrete saliva. The major functions of saliva are to **moisten** and **lubricate** the mouth at rest, but particularly during eating and speech, to **dissolve** food molecules so that they can react with gustatory receptors giving rise to the sensation of taste, to **ease swallowing**, to begin the early part of **digestion** of polysaccharides (complex sugars) and to **protect** the oral cavity by coating the teeth with a proline-rich protein or pellicle that can serve as a protective barrier on the tooth surface. Saliva also contains immunoglobulins that have a protective role in avoiding bacterial infections.

Saliva is **hypotonic** and contains a mixture of both inorganic and organic constituents. The composition varies according to which gland is secreting and also whether it is resting or being stimulated (Fig. 37c).

The **control of salivary secretion** depends on reflex responses which, in humans, have been shown to be elicited by the stimulation of gustatory (taste) receptors and periodontal and mucosal mechanoreceptors during mastication. Although it was thought that olfactory afferent stimulation (smell) also had a general reflex effect on salivary secretion, it has now been shown that this reflex operates via the submandibular/sublingual glands and not the parotid in humans. The sight and thought of food in humans have very little effect on salivary production. The perception of an increased salivary production is thought to be related to the sudden awareness of saliva already present in the mouth.

Swallowing

Swallowing occurs in a number of phases. The first phase is **voluntary** and involves the formation of a bolus of food by chewing and tongue movements (backwards and upwards), which push the food into the pharynx. The remaining phases are not voluntary, but **reflex responses** initiated by the stimulation of mechanoreceptors with afferents in the **glossopharyngeal (IX) and vagus (X) nerves** to the medulla and pons (brain stem); here, there is a group of neurones (the 'swallowing centre') which coordinates the complex sequence of events that eventually delivers the bolus into the oesophagus. The **soft palate** elevates to prevent food from entering the **nasal cavity**, respiration is inhibited, the **larynx** is raised, the **glottis** is closed and the food pushes the tip of the **epiglottis** over the tracheal opening, preventing food from entering the trachea. As the bolus enters the **oesophagus**, these changes reverse, the larynx opens and breathing continues (Fig. 37d).

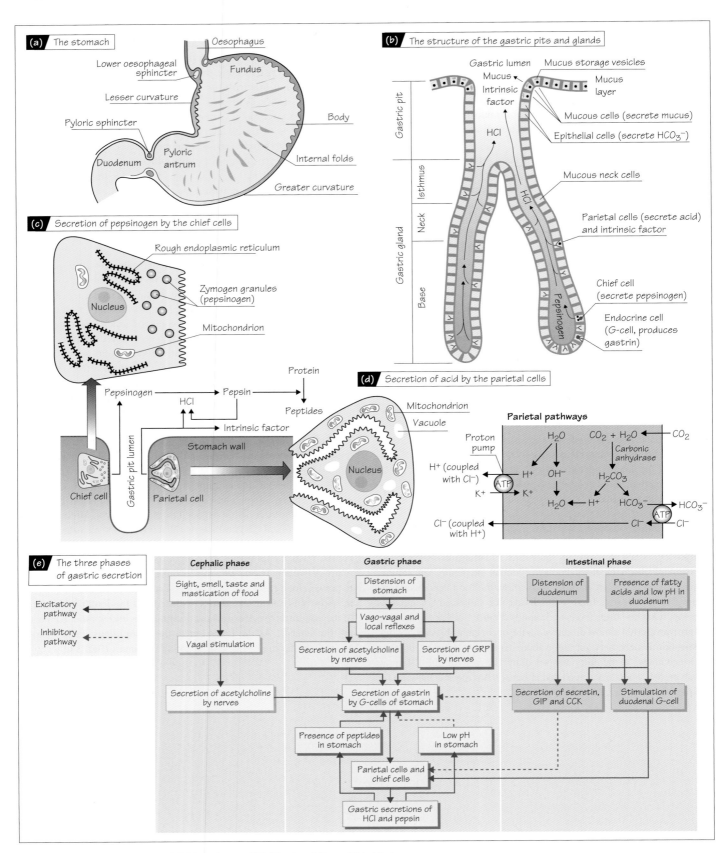

(a) The stomach

Oesophagus
Lower oesophageal sphincter
Fundus
Lesser curvature
Pyloric sphincter
Body
Pyloric antrum
Duodenum
Internal folds
Greater curvature

(b) The structure of the gastric pits and glands

Gastric lumen
Mucus storage vesicles
Mucus
Mucus layer
Intrinsic factor
Mucous cells (secrete mucus)
HCl
Epithelial cells (secrete HCO_3^-)
Gastric pit
Isthmus
Mucous neck cells
Neck
Parietal cells (secrete acid) and intrinsic factor
Gastric gland
Base
Pepsinogen
Chief cell (secrete pepsinogen)
Endocrine cell (G-cell, produces gastrin)

(c) Secretion of pepsinogen by the chief cells

Rough endoplasmic reticulum
Zymogen granules (pepsinogen)
Nucleus
Mitochondrion
Protein
Pepsinogen → Pepsin → Peptides
HCl
Intrinsic factor
Gastric pit lumen
Stomach wall
Chief cell
Parietal cell

(d) Secretion of acid by the parietal cells

Mitochondrion
Vacuole
Nucleus

Parietal pathways

Proton pump
H_2O
$CO_2 + H_2O$
CO_2
Carbonic anhydrase
H^+ (coupled with Cl^-)
H^+
OH^-
H_2CO_3
K^+
ATP
K^+
H_2O
H^+
HCO_3^-
HCO_3^-
Cl^- (coupled with H^+)
ATP
Cl^-
Cl^-

(e) The three phases of gastric secretion

Excitatory pathway ⟵
Inhibitory pathway ⟵- -

Cephalic phase	Gastric phase	Intestinal phase
Sight, smell, taste and mastication of food	Distension of stomach	Distension of duodenum — Presence of fatty acids and low pH in duodenum
Vagal stimulation	Vago-vagal and local reflexes	
	Secretion of acetylcholine by nerves — Secretion of GRP by nerves	
Secretion of acetylcholine by nerves	Secretion of gastrin by G-cells of stomach	Secretion of secretin, GIP and CCK — Stimulation of duodenal G-cell
	Presence of peptides in stomach — Low pH in stomach	
	Parietal cells and chief cells	
	Gastric secretions of HCl and pepsin	

Physiology at a Glance, Third Edition. Jeremy P.T. Ward and Roger W.A. Linden.
 © 2013 John Wiley & Sons, Ltd. Published 2013 by John Wiley & Sons, Ltd.

It is possible to swallow food and drink and for it to enter the stomach while standing on one's head or experiencing zero gravity. A ring of skeletal muscle called the **upper oesophageal sphincter** usually closes the **pharyngeal end** of the **oesophagus**. During the **oesophageal phase of swallowing**, this sphincter is relaxed, allowing the bolus of food to pass through it. Immediately afterwards, the sphincter closes. Once in the oesophagus, the bolus is propelled the 25 cm (approximately) to the **stomach** by a process called **peristalsis**, a coordinated wave of relaxation in front of the bolus and contraction behind the bolus of the circular and longitudinal muscle layers of the oesophagus, forcing the food into the stomach in about 5 s. Before the bolus enters the stomach, it passes through another sphincter, the **lower oesophageal sphincter**, formed from a ring of smooth muscle which relaxes as the peristaltic wave reaches it. The **swallowing centres in the medulla** produce a sequence of events that lead to both efferent activity to **somatic nerves** (innervating skeletal muscle) and **autonomic nerves** (innervating smooth muscle). This sequence of events is influenced by afferent receptors in the oesophagus wall sending impulses back to the medulla. The sphincters and the peristaltic waves are principally controlled by activity in the **vagus nerve** and aided by a high degree of coordination of the activity within the **enteric nerve plexuses** within the tract itself.

Once the bolus of food passes through the lower oesophageal sphincter, it enters the **stomach** (Fig. 38a). The **main functions of the stomach** are to **store** food temporarily (as it can be ingested more rapidly than it can be digested) to chemically and mechanically **digest** food using acids, enzymes and movements, to **regulate the release** of the resulting **chyme** into the small intestine, and to secrete a substance called **intrinsic factor** which is essential for the absorption of vitamin B_{12}. The stomach lies immediately below the diaphragm and, like the rest of the gastrointestinal tract, it has longitudinal and circular muscle layers and nerve plexuses in its walls; however, within the mucosa are specialized secretory cells that line the gastric glands or pits (Fig. 38b). When empty, the stomach has a volume of approximately 50 mL; however, when fully distended, its volume can be as much as 4 L. **Proteins** in the food are broken down into **polypeptides** in the stomach by enzymes called **pepsins**. These enzymes are produced in an inactive form called **pepsinogens** by the **chief cells** in the gastric mucosa, and are converted into active pepsins by the acid environment in the stomach (Fig. 38c). The acid in the stomach is **hydrochloric acid** and is produced by a specialized group of cells called **parietal cells**. The stomach can secrete as much as 2 L of acid per day, and the concentration of H^+ ions in the stomach is estimated to be about 1 million times higher than that in the blood. This concentration of H^+ ions requires a very efficient exchange of intracellular H^+ for extracellular K^+ using energy provided by the breakdown of adenosine triphosphate (ATP). This is achieved using a protein known as the **proton pump** or the H^+–K^+ ATPase protein (Fig. 38d).

The gastric mucosa does not digest itself because it is protected by an **alkaline, mucin-rich fluid** secreted by the gastric glands, which acts as a mucosal barrier by bathing the gastric epithelial cells. In addition, local mediators, such as **prostaglandins**, are released when the mucosa is irritated, and these increase the thickness of the **mucous layer** and stimulate the production of **bicarbonate** which neutralizes the acid.

Control of gastric secretions

Gastric secretions occur in basically three phases: **cephalic, gastric** and **intestinal** (Fig. 38e). The **cephalic phase** is brought about by the **sight, smell, taste** and **mastication** of food. At this stage, there is no food in the stomach and acid secretion is stimulated by the activation of the vagus and its actions on the **enteric plexus**. Postganglionic parasympathetic fibres in the **myenteric plexus** cause the release of **acetylcholine (ACh)** and stimulate the release of gastric juices from the gastric glands. Vagal stimulation also causes the release of a hormone called gastrin from cells in the antrum of the stomach called **G-cells**. Gastrin is secreted into the bloodstream and, when it reaches the gastric glands, it stimulates the release of **acid** and **pepsinogens**. Both vagal activity and gastrin also stimulate the release of **histamine** from mast cells, which, in turn, acts on **parietal cells** to produce more acid.

When food arrives in the stomach, it stimulates the **gastric phase** of secretion of **acid, pepsinogen** and **mucus**. The main stimuli for this phase are the **distension** of the stomach and the **chemical** composition of the food. Mechanoreceptors in the stomach wall are stretched and set up local myenteric reflexes and also longer vagovagal reflexes. Both cause the release of **ACh** which stimulates the release of **gastrin, histamine** and, in turn, **acid, enzymes** and **mucus**. Stimulation of the **vagus** also releases a specific peptide, **gastrin-releasing peptide (GRP)**, which mainly acts directly on the **G-cells** to release **gastrin**. Whole proteins do not affect gastric secretions directly, but their breakdown products, such as **peptides** and **free amino acids**, do so by **directly stimulating gastrin secretion**. A low pH (more acid) in the stomach **inhibits** gastrin secretion; therefore, when the stomach is empty or after food has entered it and acid has been secreted for some time, there is an inhibition of acid production. However, when food first enters the stomach, the **pH rises** (less acid) and this leads to a **release of the inhibition** and causes a **maximum secretion of gastrin**. Thus, gastric acid secretion is **self-regulating**.

The **gastric phase** normally lasts for about 3 h and the food in the stomach is converted into a sludge-like material called **chyme**. The **chyme** enters the first part of the small intestine, the **duodenum**, through the **pyloric sphincter**. The presence of chyme in the **pyloric antrum** distends it and causes antral contractions and opening of the sphincter. The rate at which the stomach empties depends on the volume in the antrum and the fall in the pH of the chyme, both leading to an increase in emptying. However, **distension** of the duodenum, the **presence of fats** and a **decrease in pH** in the duodenal lumen all cause **an inhibition of gastric emptying**. This mechanism leads to a precise supply of chyme to the intestines at a rate appropriate for it to be digested properly.

(For a description of the intestinal phase of gastric secretion, see Chapter 39.)

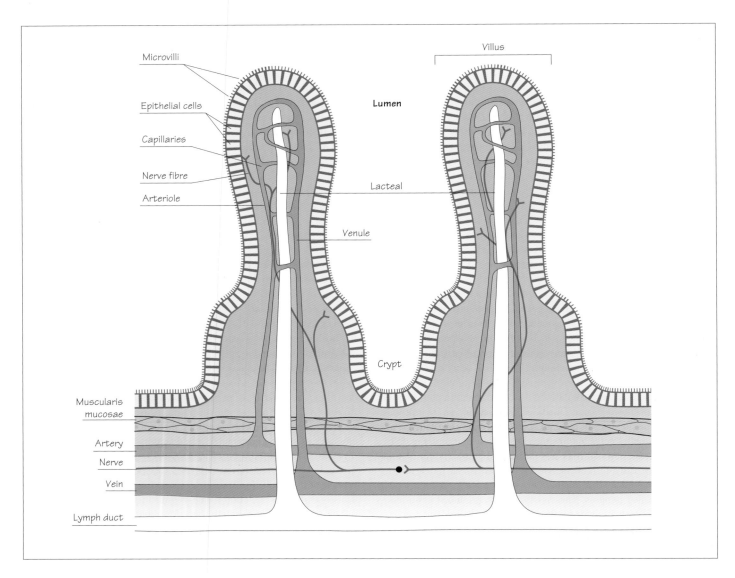

The small intestine is the main site for the digestion of food and the absorption of the products of this digestion. It is a tube, 2.5 cm in diameter and approximately 4 m in length, and comprises the **duodenum**, **jejunum** and **ileum**.

When **chyme** first enters the duodenum, there is a continuation of gastric secretion thought to be due to the activation of **G cells** in the **intestinal mucosa** (see **intestinal phase**; Fig. 38e). This is short lived as the duodenum becomes more distended with further gastric emptying. A series of reflexes is initiated which inhibits the further release of **gastric juices**. A number of **hormones** are involved in these reflex responses. **Secretin** is released in response to acid stimulation; it reaches the stomach via the bloodstream and inhibits the release of **gastrin**. The presence of **fatty acids**, due to the breakdown of fats in the duodenum itself, releases two polypeptide hormones, called **gastric inhibitory peptide** (**GIP**) and **cholecystokinin** (**CCK**), which inhibit the release of both **gastrin** and **acid**. Both secretin and CCK, however, stimulate the release of **pepsinogen** from the **chief cells**, thereby aiding protein digestion. Together with **mechanoreceptors** in the duodenum via vagal and local reflex pathways, the release of **secretin** and **CCK** has also been implicated in the control of **gastric**

Physiology at a Glance, Third Edition. Jeremy P.T. Ward and Roger W.A. Linden.
 © 2013 John Wiley & Sons, Ltd. Published 2013 by John Wiley & Sons, Ltd.

emptying. The chyme that first enters the duodenum is **acidic, hypertonic** and **only partly digested**; at this early stage, the nutrients formed cannot be absorbed. There is an **osmotic** movement of water across the freely permeable wall which leads to the contents becoming **isotonic**. The acidity is neutralized by the addition of both **bicarbonate** secreted by the **pancreas** and **bile** from the **liver**, and further digestion of the chyme is performed by the addition of enzymes from the pancreas, liver and intestine itself.

The lining of the small intestine is folded into many small, finger-like projections called **villi** (Fig. 39). Between the villi lie some small glands, called **crypts**, which can secrete up to 3 L of **hypotonic** fluid per day. The surface of the villi is covered with a layer of **epithelial cells** which, in turn, have many small projections called **microvilli** (collectively called the **brush border**) that project towards the lumen of the intestine. The small intestine is particularly adapted for the absorption of nutrients. It has a huge surface area (about the size of a tennis court), and the chyme is forced into a circular motion as it passes through the tract, facilitating mixing and therefore digestion and absorption. There is a constant turnover of epithelial cells within the gastrointestinal (GI) tract, with the small intestine epithelium totally replacing itself approximately every 6 days.

Each **villus** contains a single, blind-ended lymphatic vessel, called a **lacteal**, and also a **capillary network**. Most nutrients are absorbed into the bloodstream via these vessels. The venous drainage from the small intestine, large intestine, pancreas and also from some parts of the stomach passes via the **hepatic portal vein** into the liver; here, it passes through a second capillary bed to be further processed before returning to the circulation.

Absorption of nutrients

The small intestine absorbs **water, electrolytes, carbohydrates, amino acids, minerals, fats** and **vitamins**. The mechanisms by which movement from the lumen to the circulation occurs are variable. Nutrients move between the GI tract and the blood by passing through and around the epithelial cells. As the contents of the intestine are isotonic with body fluids and mostly have the same concentration of the major electrolytes, their absorption is active. **Water** cannot be moved directly, but follows osmotic gradients set up by the transport of ions. The major contributor to this osmotic gradient is the **sodium pump**. Na^+–K^+ **ATPase** is located on the blood side of the epithelial cell (**basolateral membrane**), and hydrolysis of adenosine triphosphate (ATP) to adenosine diphosphate (ADP) leads to the expulsion of three Na^+ ions from the cell in exchange for two K^+ ions. Both of these are against the concentration gradients, leading to a **low concentration of Na^+** and a **high concentration of K^+** within the cells. The low intracellular concentration of Na^+ ensures a movement of Na^+ from the intestinal contents into the cell by both **membrane channels** and **transported protein mechanisms**. Na^+ is then rapidly transported out of the cell again by the **basolateral Na^+–K^+ pump**. K^+ leaves the cell, again via the basolateral membrane, down its concentration gradient. This outward movement of K^+ is linked to an outward movement of Cl^-, against its concentration gradient, Cl^- having entered down its concentration gradient like Na^+ via the luminal membrane. These movements set up an osmotic gradient between the lumen and the blood, leading to water absorption following the movement of Na^+ and Cl^- from the lumen into the cell across the luminal membrane.

Carbohydrates are absorbed mostly in the form of **monosaccharides** (**glucose, fructose** and **galactose**). They are broken down into monosaccharides by enzymes released from the brush border (**maltases, isomaltases, sucrase** and **lactase**). The monosaccharides are transported across the epithelium into the bloodstream by means of cotransporter molecules that link their inward movement with that of Na^+ down its concentration gradient. At the basolateral membrane, monosaccharides leave the cell either by **simple diffusion** or by **facilitated diffusion** down the concentration gradient.

The **polypeptides** produced in the stomach are broken down into **oligopeptides** in the small intestine by enzymes (**proteases**) secreted by the pancreas: **trypsin** and **chymotrypsin**. These are further broken down into **amino acids** by another pancreatic enzyme, **carboxypeptidase**, and an enzyme located on the luminal membrane epithelial cells, **aminopeptidase**. The **free amino acids** enter the epithelial cells by secondary active transport coupled to the movement of Na^+ and a number of different cotransporter mechanisms.

Two very important minerals that are absorbed from the diet are **calcium** and **iron**. **Intracellular calcium** concentrations are low and any **free calcium** in the diet can cross the luminal membrane down a steep concentration gradient through channels or by a carrier mechanism. In the cell, it binds to a protein which carries it to the basolateral membrane, where it is actively transported against the concentration gradient by a **Ca^{2+} ATPase** with the hydrolysis of ATP, or by an **Na^+–Ca^{2+} antiporter** linked with the movement of Na^+ down its concentration gradient into the cell and the removal of Ca^{2+} from it.

Most **dietary iron** is in the **ferric** (Fe^{3+}) form which **cannot** be absorbed; however, in the **ferrous** (Fe^{2+}) form, it forms soluble complexes with **ascorbate** and other substances and **can be readily absorbed**. These complexes are transported across the membrane by a carrier protein and, once in the cell, bind with a variety of substances including **ferritin**. A second carrier protein transports the iron across the basolateral membrane into the bloodstream.

Fats and lipids

Fat digestion occurs almost entirely in the small intestine. The major enzyme is a **pancreatic enzyme** called **lipase** which breaks fat down into **monoglycerides** and **free fatty acids**. However, before the fat can be broken down, it has to be **emulsified**, which is a process by which the larger lipid droplets are broken down into much smaller droplets (about 1 μm in diameter). The main emulsifying agents are the **bile acids, cholic acid** and **chenodeoxycholic acid**. The free fatty acids and monoglycerides form tiny particles (4–5 nm in diameter) with the bile acids, called **micelles**. The outer region of the micelle is **hydrophilic** (water-attracting), whereas the inner core contains the **hydrophobic** (water-repelling) part of the molecule. This arrangement allows the micelles to enter the aqueous layers surrounding the **microvilli**, and the **monoglycerides, free fatty acids, cholesterol** and **fat-soluble vitamins** can then diffuse passively into the duodenal cells, leaving the bile salts within the lumen of the gut until they reach the ileum, where they are reabsorbed. Once within the epithelial cells, the fatty acids and monoglycerides are reassembled into fats by a number of different metabolic pathways. They then enter the lymphatic system via the **lacteals** and eventually reach the bloodstream through the **thoracic duct**.

The **fat-soluble vitamins**, A, D, E and K, essentially follow the pathways for fat absorption. The remaining **water-soluble vitamins** are mainly absorbed by diffusion or mediated transport. The exception is **vitamin B_{12}**, which must first bind with **intrinsic factor** (secreted from the parietal cells in the stomach wall). When bound, vitamin B_{12} attaches to specific sites on the epithelial cells in the **ileum** where a process of endocytosis leads to absorption.

40 The exocrine pancreas, liver and gallbladder

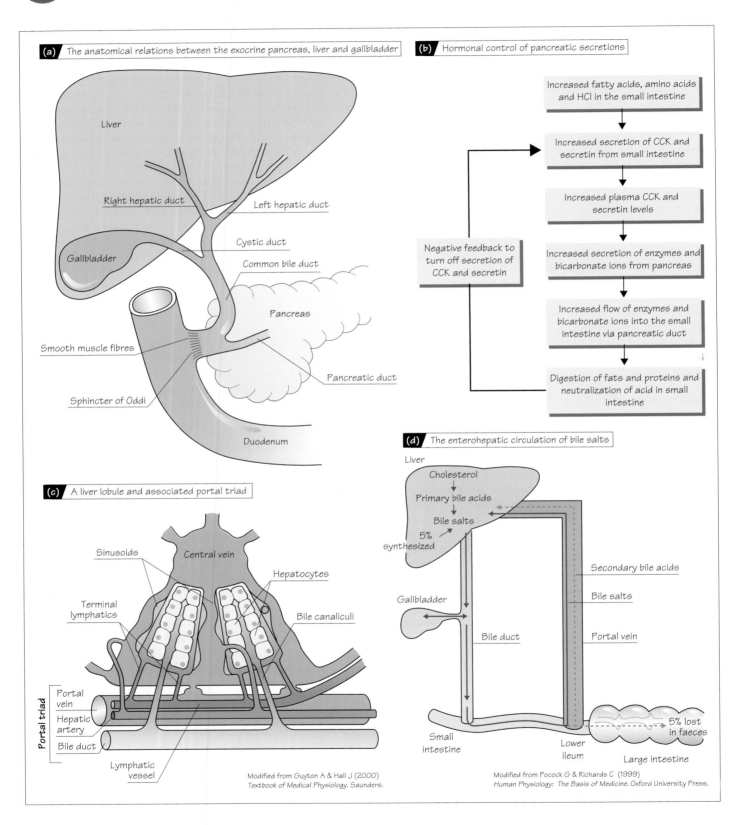

(a) The anatomical relations between the exocrine pancreas, liver and gallbladder

(b) Hormonal control of pancreatic secretions

Increased fatty acids, amino acids and HCl in the small intestine

Increased secretion of CCK and secretin from small intestine

Increased plasma CCK and secretin levels

Increased secretion of enzymes and bicarbonate ions from pancreas

Increased flow of enzymes and bicarbonate ions into the small intestine via pancreatic duct

Digestion of fats and proteins and neutralization of acid in small intestine

Negative feedback to turn off secretion of CCK and secretin

(c) A liver lobule and associated portal triad

Sinusoids, Central vein, Hepatocytes, Terminal lymphatics, Bile canaliculi, Portal triad (Portal vein, Hepatic artery, Bile duct), Lymphatic vessel

Modified from Guyton A & Hall J (2000) Textbook of Medical Physiology. Saunders.

(d) The enterohepatic circulation of bile salts

Liver, Cholesterol, Primary bile acids, Bile salts, 5% synthesized, Gallbladder, Bile duct, Secondary bile acids, Bile salts, Portal vein, Small intestine, Lower ileum, Large intestine, 5% lost in faeces

Modified from Pocock G & Richards C (1999) Human Physiology: The Basis of Medicine. Oxford University Press.

Physiology at a Glance, Third Edition. Jeremy P.T. Ward and Roger W.A. Linden.

90 © 2013 John Wiley & Sons, Ltd. Published 2013 by John Wiley & Sons, Ltd.

The pancreas

The **exocrine pancreas** secretes a major digestive fluid called **pancreatic juice**. This juice is secreted into the duodenum via the pancreatic duct that opens into the gastrointestinal (GI) tract at the same site as the **common bile duct** (see later). When food is present in the duodenum, a small sphincter (**sphincter of Oddi**) relaxes, allowing both bile and pancreatic secretions to enter the tract (Fig. 40a).

Pancreatic juice is made up of a number of **enzymes**, secreted by the **acinar** cells of the pancreas, which break down the major constituents in the diet. The enzymes include **pancreatic amylase**, which breaks down carbohydrates to monosaccharides; **pancreatic lipase**, which breaks down fats to glycerol and fatty acids; **ribonuclease** and **deoxyribonuclease**, which are involved in the breakdown of nucleic acids and free mononucleotides; and a variety of **proteolytic enzymes** (**trypsin**, **chymotrypsin**, **elastase** and **carboxypeptidase**), which break down proteins into small peptides and amino acids. The hormone **cholecystokinin** (**CCK**), released into the bloodstream by the duodenal cells in response to the presence of amino acids and fatty acids in the chyme, is responsible for the secretion of the pancreatic enzymes from the acinar cells of the pancreas. The other major secretions, besides the enzymes, are **water** and **bicarbonate ions**. The volume of pancreatic juice secreted precisely neutralizes the acid content of the chyme delivered by the stomach to the intestines. This is caused by the acid in the duodenum releasing **secretin** from its walls into the bloodstream. **Secretin** stimulates the production of water and bicarbonate ions from the duct system and, in particular, from the **epithelial cells** lining the duct. Approximately 1 L of pancreatic juice is secreted per day from a normal individual (Fig. 40b).

The liver

The **liver** is the largest organ of the body, weighing over 1 kg in the normal adult. The **functions of the liver** can be divided into two broad categories. First, it is involved with the processing of absorbed substances, both nutrient and toxic. In other words, it is responsible for the **metabolism** of a vast range of substances produced by the digestion and absorption of food from the intestine. Second, it has an important **exocrine function** in that it is involved in: (i) the production of bile acids and alkaline fluids used in the digestion and absorption of fats and for the neutralization of gastric acid in the intestines; (ii) the breakdown and production of waste products following digestion; (iii) the detoxification of noxious substances; and (iv) the excretion of waste products and the detoxification of substances in bile.

The majority of waste metabolites and detoxified substances are excreted from the body in the bile, from the GI tract, or via secretions from the liver into the bloodstream for subsequent excretion by the kidney. The relationship between the liver, gallbladder and duodenum is shown in Figure 40a. The **liver** consists of four lobes, with each lobe made up of tens of thousands of hexagonal **lobules**, 1–2 mm in diameter, which are the functional unit of the liver. Each lobule (Fig. 40c) consists of a **central vein** that eventually becomes part of the **hepatic vein**. Surrounding the central vein are single columns of liver cells (**hepatocytes**) radiating outwards; between the hepatocytes are small **canaliculi** which begin as blind-ended structures at the end nearer the central vein, but drain into the **bile duct** on the periphery of the lobule. At each of the six corners of the lobules lies a 'portal triad' comprising branches of the **hepatic artery**, the **portal vein** and the **bile duct**. The bile ducts eventually drain into the **terminal bile duct**.

Bile and the gallbladder

The **hepatocytes** secrete a fluid called **hepatic bile**. It is isotonic and resembles plasma ionically. It also contains **bile salts**, **bile pigments**, **cholesterol**, **lecithin** and **mucus**. This fraction of bile is called the **bile acid-dependent fraction**. As it passes along the bile duct, the bile is modified by the epithelial cells lining the duct by the addition of **water** and **bicarbonate ions**; this fraction is called the **bile acid-independent fraction**. Overall, the liver can produce 500–1000 mL of bile per day. The bile is either discharged directly into the duodenum or stored in the **gallbladder**. The bile acid-independent fraction is made at the time it is required, i.e. during digestion of the chyme. The bile acid-dependent fraction is made when the bile salts are returned from the GI tract to the liver, and is then stored in the gallbladder when the sphincter of Oddi is closed. About 95% of the bile salts that enter the small intestine in bile are recycled and reabsorbed into the portal circulation by active transport mechanisms in the distal ileum (the so-called **enterohepatic circulation**; Fig. 40d). Many of the bile salts are returned unaltered, some are broken down by intestinal bacteria into **secondary bile acids** and then reabsorbed, and a small proportion escapes reabsorption and is excreted in the faeces.

The **gallbladder** not only stores the bile, but also concentrates it by removing non-essential solutes and water, leaving the bile acids and pigments. The process of concentration is mainly by active transport of Na^+ ions into the intercellular spaces of the lining cells and this, in turn, draws water, HCO_3^- and Cl^- ions from the bile and into the extracellular fluid, thereby concentrating the bile held in the gallbladder.

The formation of bile is stimulated by **bile salts**, **secretin**, **glucagons** and **gastrin**. The release of bile stored in the gallbladder, however, is stimulated by the secretion of **CCK** into the bloodstream when chyme enters the duodenum and, to a lesser extent, by the actions of the **vagus nerve**. Within a few minutes of a meal, particularly when fats are consumed, the muscles of the gallbladder contract; this forces the contents into the duodenum through the now relaxed sphincter of Oddi. CCK relaxes the sphincter and stimulates the pancreatic secretions at the same time. The gallbladder empties completely 1 h after a fat-rich meal and maintains the level of bile acids in the duodenum above that necessary for the function of the micelles.

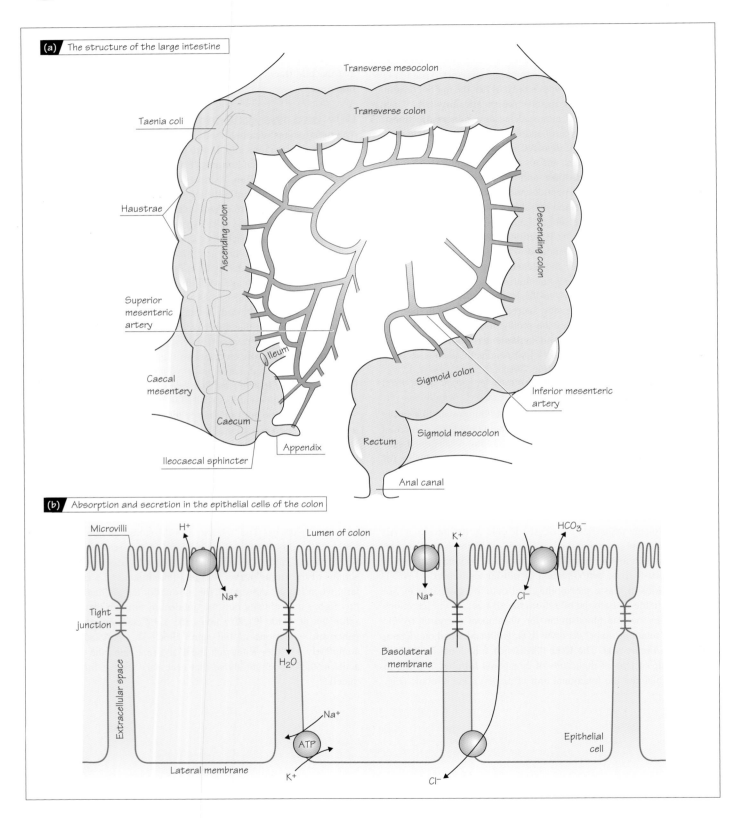

(a) The structure of the large intestine

Transverse mesocolon

Transverse colon

Taenia coli

Haustrae

Ascending colon

Descending colon

Superior mesenteric artery

Ileum

Caecal mesentery

Sigmoid colon

Inferior mesenteric artery

Caecum

Rectum

Sigmoid mesocolon

Appendix

Ileocaecal sphincter

Anal canal

(b) Absorption and secretion in the epithelial cells of the colon

Microvilli

H^+

Lumen of colon

K^+

HCO_3^-

Na^+

Na^+

Cl^-

Tight junction

Extracellular space

H_2O

Basolateral membrane

Na^+

ATP

Cl^-

Epithelial cell

Lateral membrane

K^+

Cl^-

Physiology at a Glance, Third Edition. Jeremy P.T. Ward and Roger W.A. Linden.
© 2013 John Wiley & Sons, Ltd. Published 2013 by John Wiley & Sons, Ltd.

The **large intestine** comprises the **caecum, ascending, transverse, descending** and **sigmoid colon, rectum** and **anal canal** (Fig. 41a). It is approximately 1.2 m in length and between 6 and 9 cm in diameter. Approximately 1.5 L of chyme enters the large intestine per day through a sphincter called the **ileocaecal sphincter**. Distension of the terminal ileum results in the opening of the sphincter and distension of the caecum causes it to close, thereby maintaining the optimum rate of entry to maximize the main function of the large intestine, which is to absorb most of the water and electrolytes. The initial 1.5 L is reduced to about 150 g of **faeces** consisting of 100 mL of water and 50 g of solids.

The muscle layers of the large intestine are slightly different from those found in the rest of the gastrointestinal (GI) tract. It still has a powerful circular muscle layer, but its **longitudinal muscle layer** is concentrated into **three bands** called the **taeniae coli**. The caecum and the ascending and transverse colon are innervated by **parasympathetic** branches of the **vagus**; the descending and sigmoid colon, rectum and anal canal are innervated by **parasympathetic** branches of the **pelvic nerves** from the sacral spinal cord. These parasympathetic fibres innervate the intramural plexuses. The **sympathetic** nerves via the superior mesenteric plexus, and via the inferior mesenteric and the superior hypogastric plexuses, innervate the proximal and distal parts of the large intestine, respectively. The rectum and anal canal are innervated via the inferior hypogastric plexus. Stimulation of the parasympathetic fibres causes segmental contraction, whereas stimulation of the sympathetic fibres stops colonic activity. The **internal** and **external anal sphincters** usually keep the anal canal closed and are controlled both **reflexly** and **voluntarily**. The **internal sphincter** is made up of **circular smooth muscle**, and the more distal **external sphincter** is composed of **striated muscle** which is innervated by motor fibres from the **pudendal nerve**.

Movement of the chyme through the large intestine involves both **mixing** and **propulsion**. However, as the main function is to store the residues of food and to absorb water and electrolytes from it, the movements are slow and sluggish (approximately 5–10 cm/h). Chyme usually remains in the colon for up to 20 h. The mixing movement is called **haustration** and the sac-like compartments in the colon caused by this process are called **haustra**. The contents of the haustra are often shunted back and forth from one to another in a process called **haustral shuttling**. This aids the exposure of chyme to the mucosal surface and helps the reabsorption of water and electrolytes. In the distal parts of the colon, the contractions are slower and less propulsive, and eventually the faeces collect in the descending colon.

Several times a day there is an increase in activity within the colon, in which there is a vigorous propulsive movement, the **mass movement**. This results in the emptying of a large proportion of the content of the proximal colon into the more distal parts. This **mass movement** is initiated by a complex series of intrinsic reflex pathways started by the distension of the stomach and duodenum soon after the consumption of a meal.

Defecation

When a critical mass of faeces is forced into the rectum, the desire for **defecation** is experienced. This sudden distension of the rectum walls produced by the final mass movement leads to a **defecation reflex**. This reflex comprises a contraction of the rectum, relaxation of the internal anal sphincter and, initially, contraction of the external anal sphincter. This initial contraction is soon followed by a reflex relaxation of the sphincter initiated by an increase in the peristaltic activity in the sigmoid colon and pressure in the rectum. The faeces are then expelled. This reflex relaxation can be overridden by higher centre activity, leading to a voluntary control over the sphincter which can delay the expulsion of faeces. The prolonged distension of the rectum then leads to a **reverse peristalsis**, which empties the rectum into the colon and removes the urge to defecate until the next mass movement and/or a more convenient time.

The chyme that enters the large intestine is **isotonic**; however, in the colon more water than electrolytes is absorbed, leading to water being absorbed against a concentration gradient. The process is controlled by **Na$^+$–K$^+$ ATPases** located in the basolateral and lateral membranes of the epithelial cells that line the walls (Fig. 41b). The mucosal surface of the large intestine is relatively smooth with no villi (only microvilli); however, crypts are present and the majority of cells are **columnar** absorptive cells with a large number of mucous-secreting goblet cells. Na$^+$ is extruded by the membrane pumps into the extracellular spaces. **Tight junctions** at the luminal side of the cells prevent the diffusion of Na$^+$ and Cl$^-$ from the extracellular spaces into the lumen; this leaves a hypertonic solution close to the lumen, causing water to diffuse from the contents of the lumen. The electrolytes are absorbed by a variety of mechanisms similar to those described for the small intestine. Essentially, there is a net movement of K$^+$ and HCO$_3^-$ ions from the blood into the large intestine because of the potential difference set up by the asymmetrical absorption of Na$^+$ and Cl$^-$ across the cell wall.

Gut microflora

Most of the **bacteria** that are present in the GI tract are found in the large intestine, because the acid environment in the rest of the tract destroys most of the so-called microflora. Ninety-nine per cent of the bacteria are anaerobic and most are lost in the faeces (which is said to contain 10^{11} bacteria per gram). The **bacteria** are involved in the **synthesis** of **vitamins K, B$_{12}$, thiamine** and **riboflavin**, the **breakdown** of **primary** to **secondary bile acids** and the **conversion** of **bilirubin** to **non-pigmented metabolites**, all of which are readily absorbed by the GI tract. The bacteria also break down cholesterol, some food additives and drugs.

Table 42 Tissues involved in endocrine control systems. The top half of the table shows the products of classical glandular tissue; the bottom half lists some of the other organs that release hormones

Secreting tissue	Main hormone(s)	Main target tissue(s)	Discussed in chapter
Glands			
Anterior pituitary (P)	Adrenocorticotrophic hormone (ACTH)	Adrenal cortex (M)	49
	Growth hormone (GH)	Liver, bones, muscle (G)	47
	Follicle-stimulating hormone (FSH)	Gonads (R)	
	Luteinizing hormone (LH)	Gonads (R)	50
	Prolactin	Mammary glands (R)	53
	Thyroid-stimulating hormone (TSH)	Thyroid gland (G, M)	45
Intermediate pituitary (P)	Melanotrophin-stimulating hormone (MSH)	Melanocytes (H)	44
Posterior pituitary (P)	Antidiuretic hormone (ADH)	Kidney (H)	35
	Oxytocin	Mammary glands	53
		Uterus (R)	52
Pineal (A)	Melatonin	Hypothalamus (H)	
Thyroid (A)	Thyroxine (T4)	Most tissues (G, M)	46
	Tri-iodothyronine (T3)	Most tissues (G, M)	
	Calcitonin	Bones, gut (H)	48
Parathyroid (P)	Parathyroid hormone (PTH)	Bones, gut (H)	48
Pancreas (P)	Insulin	Liver, muscle, adipose tissue (G, M, H)	43
	Glucagon		
Adrenal cortex (S)	Corticosteroids (including cortisol)	Multiple (G, M)	49
	Aldosterone	Kidney (H)	35 and 49
Adrenal medulla (A)	Adrenaline (epinephrine)	Multiple (H, M)	49
	Noradrenaline (norepinephrine)	Multiple (H, M)	
Gonads: male (S)	Testosterone	Testes (R)	50
Gonads: female (S)	Oestradiol	Ovaries, uterus (R)	50 and 52
	Progesterone	Ovaries, uterus (R)	
	Human chorionic gonadotrophin (hCG)	Uterus (R)	52
Placenta (P, S)	Oestradiol	Ovaries, uterus (R)	
	Progesterone	Ovaries, uterus (R)	
Non-glands			
Brain (P, A)	Hypothalamic-releasing hormones	Anterior pituitary gland (H, R, M)	44
	Growth factors	Various (M)	
Heart (P)	Atrial natriuretic peptide (ANP)	Kidney	35
Kidney (P, S)	Erythropoietin (EPO)	Bone marrow (M)	8
	1,25-Dihydroxycholecalciferol	Gut, kidney (H)	48
	Renin	Plasma proteins (H)	35
Liver (P)	Insulin-like growth factor-1 (IGF-1)	Various (M)	47

Physiology at a Glance, Third Edition. Jeremy P.T. Ward and Roger W.A. Linden.

 © 2013 John Wiley & Sons, Ltd. Published 2013 by John Wiley & Sons, Ltd.

Table 42 (*Continued*)

Secreting tissue	Main hormone(s)	Main target tissue(s)	Discussed in chapter
Adipose tissue (P)	Leptin	Hypothalamus (M)	44
Gastrointestinal tract (P, A)	Gastrin	Gut (H, M)	37–41
	Secretin		
	Cholecystokinin (CCK)		
	Vasoactive intestinal polypeptide (VIP)		
	Gastrin-releasing peptide (GRP)		
Immune cells (P)	Cytokines	Hypothalamus (H, M)	10
	Opioid peptides		
Platelets (P)	Growth factors	Various (G)	46
Various sites (P)	Growth factors	Various (G)	46
	Neurotrophins	Neurones (G)	

Molecules: A, modified amino acid; P, peptide/protein; S, steroids/sterols.
Functions: H, homeostasis; R, reproduction; G, growth and development; M, metabolism.

Multicellular organisms must coordinate the diverse activities of their cells, often over large distances. In animals, such coordination is achieved by the nervous and **endocrine** systems, the former providing rapid, precise but short-term control and the latter providing generally slower and more sustained signals. The two systems are intimately integrated and in some places difficult to differentiate. Endocrine control is mediated by **hormones**, signal molecules usually secreted in low concentrations (10^{-12}–10^{-7} M) into the bloodstream, so that they can reach all parts of the body. Other types of chemical communication are mediated over smaller distances. Chemical signals can act locally on neighbouring cells (paracrine signals) or can act on the same cell that produced the signal (autocrine signals); **juxtacrine** communication requires direct physical contact between signal chemicals on the surface of one cell and receptor molecules on the surface of a neighbour. Many hormones are secreted by discrete glands (Table 42), while others are released from tissues with other primary functions. For instance, several of the cytokines released by immune cells (Chapter 10) act at some distance from their site of release and can fairly be considered as hormones.

Features of hormonal signalling

Hormonal molecules can be: (i) **modified amino acids** [e.g. adrenaline (norepinephrine); Chapter 49]; (ii) **peptides** (e.g. somatostatin; Chapter 44); (iii) **proteins** (e.g. insulin; Chapter 43); or (iv) derivatives of the fatty acid cholesterol, such as **steroids** (e.g. cortisol; Chapter 49; Table 42). Protein and peptide hormones are cleaved from larger gene products, whereas smaller molecules require the precursor to be transported into endocrine cells so that it can be modified by sequences of enzymes to generate the final product (e.g. Chapter 49). Most hormones are stored in intracellular membrane-bound **secretory granules**, to be released by a **calcium-dependent mechanism** similar to the release of neurotransmitters from nerve cells (Chapter 7; Fig. 43b) when the cell is activated. However, thyroid hormones and steroids, which are highly lipid soluble, cannot be stored in this way. Most steroids are made immediately before release, whereas the thyroid hormones are bound within a glycoprotein matrix (Chapter 45). After secretion, some hormones bind to **plasma proteins**. In most cases this involves non-specific binding to albumin, but there are **specific binding proteins** for some hormones, such as cortisol or testosterone. A hormone bound to a plasma protein cannot reach its site of action and is protected from metabolic degradation, but is freed when the plasma level of the hormone falls. The bound fraction thus acts as a reservoir that helps to maintain steady plasma levels of the free hormone.

Hormones exert their effects by interactions with **specific receptor proteins** and will act only on cells carrying those receptors. Most protein and peptide hormones activate cell surface receptors that are coupled to guanosine triphosphate-binding proteins (**G-proteins**) (Chapter 4) or that have **intrinsic tyrosine kinase** activity (e.g. Chapter 43). Receptors for lipid-soluble hormones (steroids, thyroid hormones) are usually *inside* the target cell, and modify gene transcription directly (e.g. Chapter 45). Because they are in the bloodstream, free hormones can reach all of the tissues that bear the appropriate receptors. Endocrine signals therefore provide a good way of inducing simultaneous changes in multiple organs, and most hormones have effects in more than one tissue. A corollary of this position is that many physiological processes are influenced by more than one hormone, as will become clear in subsequent chapters. Hormones are **inactivated** by metabolic transformation by enzymes, usually in the liver or at the site of action. It is a general rule that the smaller the hormone, the more rapid its inactivation.

Control of hormones

Endocrine secretion can be controlled by the nervous system, other endocrine glands, or respond directly to levels of metabolites in the environment of the gland, and most are subject to all of these types of control. A common feature of hormonal control systems is a heavy reliance on **negative feedback loops**. Almost all hormones feed back to inhibit their own release, providing a direct method of moderating the output of hormone into the blood (Chapters 44–53). A less common feature of endocrine systems, associated only with reproductive functions, is *positive* feedback, whereby the release of a hormone leads to events that further promote release (Chapters 50, 51 and 53). The carriage of hormones in the blood provides a limit on how quickly hormones produce their effects. The relatively slow nature of hormonal signalling puts limits on the types of physiological processes that can be controlled by hormones. They fall into four broad categories: (i) **homeostasis**; (ii) **reproduction**; (iii) **growth and development**; and (iv) **metabolism**. These systems work over time-scales that range from a few minutes (e.g. milk ejection; Chapter 53) to years (growth; Chapter 47).

LIVERPOOL JOHN MOORES UNIVERSITY
LEARNING SERVICES

43 Control of metabolic fuels

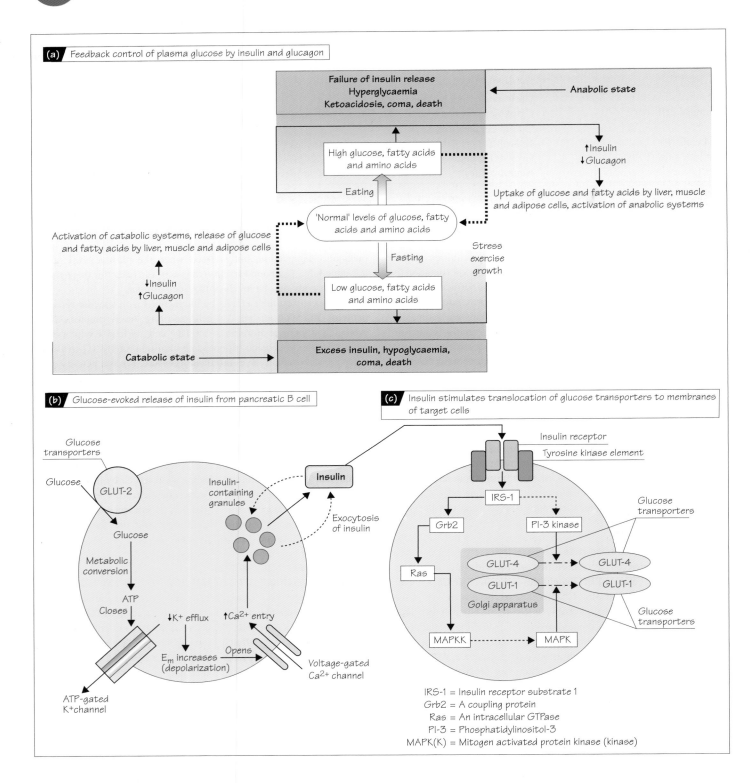

(a) Feedback control of plasma glucose by insulin and glucagon

Failure of insulin release
Hyperglycaemia
Ketoacidosis, coma, death

Anabolic state

↑Insulin
↓Glucagon

High glucose, fatty acids
and amino acids

Eating

Uptake of glucose and fatty acids by liver, muscle
and adipose cells, activation of anabolic systems

'Normal' levels of glucose, fatty
acids and amino acids

Activation of catabolic systems, release of glucose
and fatty acids by liver, muscle and adipose cells

Stress
exercise
growth

Fasting

↓Insulin
↑Glucagon

Low glucose, fatty acids
and amino acids

Catabolic state

Excess insulin, hypoglycaemia,
coma, death

(b) Glucose-evoked release of insulin from pancreatic B cell

Glucose
transporters

Glucose

GLUT-2

Glucose

Metabolic
conversion

ATP
Closes

↓K⁺ efflux

↑Ca²⁺ entry

Em increases
(depolarization)

Opens

ATP-gated
K⁺channel

Voltage-gated
Ca²⁺ channel

Insulin-
containing
granules

Insulin

Exocytosis
of insulin

(c) Insulin stimulates translocation of glucose transporters to membranes
of target cells

Insulin receptor
Tyrosine kinase element

IRS-1

Grb2

PI-3 kinase

Glucose
transporters

Ras

GLUT-4

GLUT-1

Golgi apparatus

GLUT-4

GLUT-1

Glucose
transporters

MAPKK

MAPK

Glucose
transporters

IRS-1 = Insulin receptor substrate 1
Grb2 = A coupling protein
Ras = An intracellular GTPase
PI-3 = Phosphatidylinositol-3
MAPK(K) = Mitogen activated protein kinase (kinase)

Physiology at a Glance, Third Edition. Jeremy P.T. Ward and Roger W.A. Linden.

 © 2013 John Wiley & Sons, Ltd. Published 2013 by John Wiley & Sons, Ltd.

Animal cells utilize glucose and fatty acids as fuels to generate the energy-rich molecule adenosine triphosphate (ATP) (Chapter 3). The blood levels of these molecules must be carefully controlled to ensure a steady supply of fuel to active tissues, a task that is complicated by the tendency of many animals (not ruminants) to eat discrete meals rather than continuously. Immediately after a meal, circulating levels of fuel molecules rise and any excess to immediate requirements is stored. This requires the transport of the molecules into cells (primarily **liver**, **skeletal muscle** and the fat-storing cells of **adipose tissues**) and the synthesis of storage molecules, such as **glycogen**, a polymer of glucose, **triglycerides** (fats) and, to a lesser extent, proteins. As time after a meal increases, the consumption of blood glucose and fatty acids necessitates the activation of tissue energy stores. Glycogen is broken down into glucose, triglycerides are converted into free fatty acids and ketone bodies and, if the fast is prolonged, proteins are catabolized to provide a supply of amino acids that can be converted to glucose (**gluconeogenesis**). The body thus alternates between two states, which can be described as **anabolic**, in which storage molecules are manufactured, and **catabolic** in which the same molecules are broken down (Fig. 43a). Switching between these states is controlled mainly by hormones, with the pancreatic proteins **insulin** and **glucagon** being the prime movers of the anabolic and catabolic processes, respectively. In addition, growth hormone (Chapter 47), cortisol, adrenaline (epinephrine) and noradrenaline (norepinephrine) (Chapter 49) can stimulate catabolic processes (Fig. 43a). There is growing evidence that hormones produced from fat (e.g. leptin) and the gut (e.g. ghrelin from the stomach) are involved in energy homeostasis, including controlling food intake, energy expenditure and adiposity.

Insulin and glucagon

These hormones are made in the endocrine tissues of the pancreas, known as the **islets of Langerhans**. Three main types of cell have been identified within the islets: peripherally located **A** (also known as α) cells, which manufacture and secrete **glucagon**; centrally located **B** (or β) cells for the production and release of **insulin**; and **D** (δ) cells that synthesize and liberate **somatostatin**. The exact role of somatostatin has not been established, but it may be involved in controlling the release of the other two hormones. Insulin release is stimulated initially during eating by the parasympathetic nervous system and gut hormones, such as secretin (Chapter 39), but most output is driven by the rise in plasma glucose concentration that occurs after a meal (Fig. 43a,b). Circulating fatty acids, ketone bodies and amino acids augment the effect of glucose. The major action of insulin is to stimulate glucose uptake, with the subsequent manufacture of glycogen and triglycerides by adipose, muscle and liver cells. Its effects are mediated by a **receptor tyrosine kinase** (RTK; Fig. 43c; Chapter 47). The enzyme activates an intracellular pathway that results in the translocation of the glucose transporter GLUT-4 and to a lesser extent GLUT-1 to the plasma membrane of the affected cell, to facilitate the entry of glucose (Fig. 43c). Insulin thus decreases plasma glucose. Insulin release is reduced as the blood glucose concentration falls, and is further inhibited by catecholamines (Chapter 49) acting at B-cell α_2-adrenoceptors (Chapters 7 and 49). Glucagon release patterns tend to be the mirror image of those of insulin. Low blood glucose initiates glucagon release directly and also drives nervous and hormonal release

of catecholamines, which activate β-adrenoceptors (Chapters 7 and 49) on A cells to augment glucagon release. Glucagon acts on guanosine triphosphate-binding protein (G-protein)-coupled receptors that stimulate the production of intracellular cyclic adenosine monophosphate (cAMP) (Chapter 4). In liver cells, this results in the inhibition of glycogen synthesis and the activation of glycogen breakdown systems. Similar effects are obtained in muscle cells to increase circulating levels of glucose. There are interactions between glucagon and insulin within the islets: insulin inhibits A-cell release of glucagon, but glucagon *stimulates* the release of insulin, an effect that ensures a basal level of insulin release irrespective of glucose levels. The two hormones operate as part of a classical negative feedback system (Fig. 43a; Chapter 1), in which the A and B cells act as combined sensors–comparators, and their hormones activate the effector tissues.

Diabetes mellitus

This disease is caused by failure of B-cell function, either by **autoimmune attack**, in which the immune system (Chapter 10) misidentifies the cells as non-self and destroys them, or by pathologies, such as **obesity**, that impair insulin release. The former type of disease is usually early onset and is treated with insulin (**insulin-dependent diabetes**), whereas the latter develops later and is treated by diets that lower blood glucose levels or drugs that stimulate insulin release (**non–insulin-dependent diabetes**). Untreated, the condition leads to chronically high levels of plasma glucose (**hyperglycaemia**), overloading of the kidney glucose transporters (Chapter 33) so that sugar begins to appear in the urine. The osmotic effect of glucose leads to excess production of urine (**polyuria**) that tastes sweet (this used to be the diagnostic test for diabetes, and gives the disease its name; Latin *mellitus* = sweet). Long-term hyperglycaemia drives excessive lipolysis by liver cells, leading to a build-up of ketone bodies and the condition known as **ketoacidosis**. This disrupts brain function, causing coma and eventually death. A sharp fall in blood glucose (**hypoglycaemia**) caused by excessive insulin administration starves the brain of its main metabolic fuel and, by a sad irony, can also lead to coma and death (Fig. 43a). The main symptoms of hyper- and hypoglycaemia are shown in Table 43. Long-term complications of diabetes include damage to small blood vessels, especially in the retina and renal nephron (diabetic retinopathy and nephropathy). This is at least partly due oxidative stress as a result of the hyperglycaemia.

Table 43 The main symptoms of hyper- and hypo-glycaemia

	Hyperglycaemia	Hypoglycaemia
	Polydypsia (excessive thirst)	Increased appetite
	Polyuria	Headache
	Fatigue	Weakness
	Slow healing of wounds	Loss of concentration
	Impaired vision	Blurred vision
Extreme	Ketone breath	Coma
	Coma	

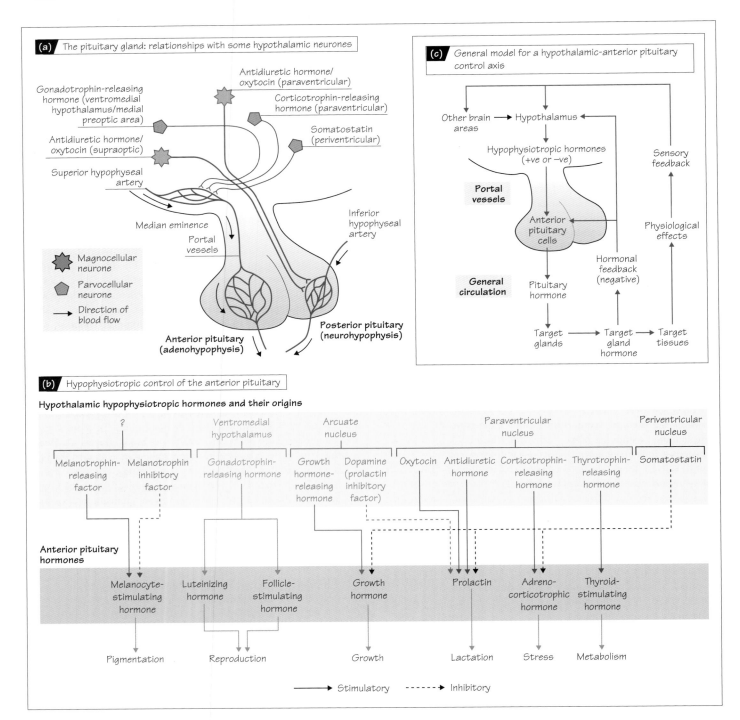

Physiology at a Glance, Third Edition. Jeremy P.T. Ward and Roger W.A. Linden.

© 2013 John Wiley & Sons, Ltd. Published 2013 by John Wiley & Sons, Ltd.

The **pituitary gland**, which is under the direct control of the brain from the **hypothalamus**, provides endocrine control of many major physiological functions. The hypothalamus is composed of a number of nuclei (collections of cell bodies) and vaguely defined 'areas', and surrounds the **third ventricle** at the base of the medial forebrain. The most important hypothalamic areas for endocrine function are the **paraventricular**, **periventricular**, **supraoptic** and **arcuate nuclei**, and the **ventromedial hypothalamus**. Some of the hypothalamic neurones can secrete hormones (a process called *neurosecretion*), releasing chemicals in exactly the same way as other nerve cells (Chapter 7), albeit that their signals are liberated into the bloodstream rather than into synapses (Fig. 44a) The pituitary is located immediately beneath the hypothalamus and comprises three divisions: the **anterior pituitary**, the **intermediate lobe** (almost vestigial in humans) and the **posterior pituitary** (Fig. 44a,b). The anterior pituitary develops from tissues originating in the roof of the mouth, is non-neural and is sometimes known as the **adenohypophysis**. The posterior gland is really an extension from the hypothalamus itself, consists of neural tissue and is referred to as the **neurohypophysis**. All pituitary hormones are either peptides or proteins. As befits their developmental origins, the adeno- and neuro-hypophyses are controlled in different ways.

The anterior pituitary and intermediate lobe

The adenohypophyseal hormones and their actions are listed in Figure 44b. They are released under the control of chemical signals (**hypothalamic releasing or inhibiting hormones**) originating from small (**parvocellular**) neurones with their cell bodies in the hypothalamus (Fig. 44a–c). These hormones are peptides or proteins released into the blood at the **median eminence** (Fig. 44a) when the appropriate parvocellular neurones are electrically active. The hypothalamic hormones are transported directly to the anterior pituitary via the **hypophyseal portal vessels** (Fig. 44a). The portal vessels carry hypophysiotropic signals directly to the anterior pituitary to stimulate *or* inhibit the release of pituitary hormones by the activation of receptors on specific groups of pituitary cells (Fig. 44b). It should be noted that some hypothalamic hormones control more than one pituitary hormone. Figure 44c illustrates the basic principles that underlie the control of anterior pituitary hormones; this is a form of chemical cascade that allows for the precise control of pituitary output with two stages of

signal amplification: first, at the pituitary itself, where tiny amounts of hypothalamic hormones control the release of larger quantities of pituitary hormone; and then at the final target gland, where the pituitary signals stimulate the release of still larger quantities of hormones such as steroids. The cascade allows for feedback control of hormone release at several points. The final hormone (and often some of the intermediate signals) inhibits further activity in the axis to provide the fine regulation of hormone release (Fig. 44c). This is a characteristic feature of anterior pituitary control systems.

The posterior pituitary

The posterior gland secretes two peptide hormones: **oxytocin** and **antidiuretic hormone** (**ADH**; also known as **vasopressin**). The hormones are manufactured in the cell bodies of large (**magnocellular**) neurones in the supraoptic and paraventricular nuclei of the hypothalamus, and are transported down the axons of these cells to their terminals on capillaries originating from the **inferior hypophyseal artery** within the posterior pituitary gland (Fig. 44a). When magnocellular neurones are activated (see Chapters 35, 52 and 53), they release oxytocin or ADH into the general circulation, from whence they can reach the relevant target tissues to produce the required effect. The signals that drive the release of posterior gland hormones are entirely neural, so that the hormones are said to be involved in **neuroendocrine reflexes**. These hormones operate over shorter time courses (minutes) than most endocrine events (hours to days). The release of ADH is controlled by conventional negative feedback mechanisms based on plasma osmolality and blood volume (Chapter 35). Oxytocin, however, is involved in *positive* feedback mechanisms (Chapters 52 and 53).

Pulsatile release of pituitary hormones

Hormones released from the hypothalamus tend to appear in the blood in discrete pulses, rather than as continuous secretions. This is achieved by the synchronous activation of hormone-releasing neurones of the hypothalamus. As will be seen in later chapters, episodic release has profound implications for the operation of the endocrine system. It also raises a number of interesting and as yet unanswered questions as to how many separate and more or less widely scattered neurones can be activated simultaneously to give rise to pulsatile release.

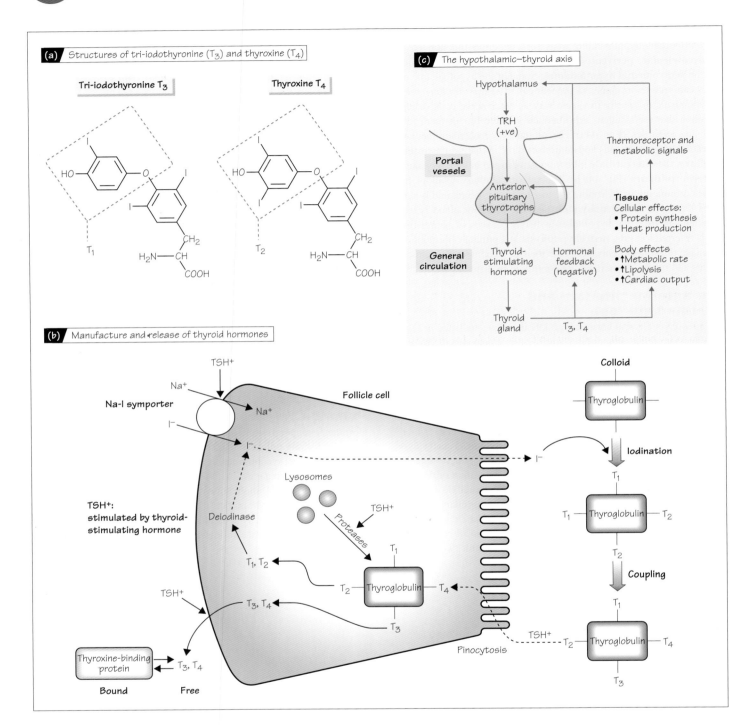

(a) Structures of tri-iodothyronine (T₃) and thyroxine (T₄)

Tri-iodothyronine T₃

Thyroxine T₄

(c) The hypothalamic–thyroid axis

(b) Manufacture and release of thyroid hormones

Physiology at a Glance, Third Edition. Jeremy P.T. Ward and Roger W.A. Linden.

 © 2013 John Wiley & Sons, Ltd. Published 2013 by John Wiley & Sons, Ltd.

The thyroid gland is attached to the anterior surface of the trachea just below the larynx. It releases two iodine-containing hormones, **thyroxine** (also known as T_4) and **tri-iodothyronine** (T_3; Fig. 45a), the main effect of which is to increase heat production (**thermogenesis**) throughout the body and thereby induce an increase in metabolic rate. The hormones also have a crucial role in growth and development.

Synthesis and release

The thyroid gland is formed from clusters of cells (**follicles**) that surround a gel-like matrix or colloid, the primary constituent of which is the glycoprotein **thyroglobulin**. The follicle cells actively accumulate iodide (I^-) ions by means of an Na^+–I^- symporter (Chapter 4) driven by the inward sodium gradient (Fig. 45b). The formation of T_3 and T_4 occurs in two steps: (i) the amino acid tyrosine is iodinated to form mono- (T_1) or di-iodotyrosine (T_2) (Fig. 45a); (ii) T_2 is then coupled to T_1 or T_2 by thyroperoxidase to form the thyroid hormones. This process occurs with the tyrosine residues attached to thyroglobulin, so that, at any one time, this protein is festooned with molecules of T_1, T_2, T_3 and T_4 (Fig. 45b). The thyroid hormones and their intermediates are highly lipophilic and would escape from the gland were they not incorporated into thyroglobulin, which thus acts as a nucleus for the manufacture of the hormones and as a storage site. The hormones are released under the control of **thyroid-stimulating hormone (TSH)** from the anterior pituitary, which is obligatory for normal thyroid function (Chapter 44; Fig. 45c). Under the action of TSH, thyroid follicle cells pinch off small quantities of colloid by **pinocytosis**. Lysozymal protease enzymes then act on the thyroglobulin to liberate the iodinated compounds into the cell and thence into the bloodstream (Fig. 45b). Free T_1 and T_2 are deiodinated by enzymatic action before they can leave the cell. The average plasma concentration of T_3 is roughly one-sixth of that of T_4, and much of that derives from deiodinated T_4. Most of the thyroid hormones in the blood are bound to thyroxine-binding protein and are thus unavailable to their receptors, which are located *inside* target cells, attached directly to deoxyribonucleic acid (DNA). The small amounts of free T_3 and T_4 in plasma readily cross the cell membranes to bind to thyroid hormone receptors (the most important of which is **$TR\alpha_1$**). Thyroid receptors are linked to a DNA sequence known as the **thyroid-response element (TRE)** which initiates the transcription of thyroid-responsive genes. T_3 is some 10 times more potent than T_4 in activating $TR\alpha_1$ and consequently mediates most thyroid hormone actions, notwithstanding its lower levels in plasma. Thyroid receptors are present in almost all tissues, with particularly high levels in the liver and low levels in the spleen and testes.

Physiological roles of thyroid hormones

Basal levels of thyroid hormone release are essential to maintain a normal metabolic rate. Situations requiring increased heat production, for instance when the core temperature falls, lead to enhanced activation of the thyroid axis. The effects take up to 4 days to reach a maximum, a slow time course that is characteristic of hormones acting through nuclear receptors. The primary action of thyroid hormones is an increase in the synthesis of Na^+–K^+ ATPase (Chapter 4), an enzyme that consumes large amounts of metabolic energy, to increase heat production. The hormones may also enhance the production of **uncoupling proteins (UCPs)**. These molecules act in mitochondria to divert the H^+ ion gradient generated by the electron transport chain (Chapter 4), so that it produces heat rather than driving adenosine triphosphate (ATP) synthase. Although UCP-1 is found only in brown fat, a tissue that is uncommon in adult humans, two other members of the family (UCP-2 and UCP-3) are present in muscle and other tissues, and may thus contribute to thyroid-stimulated thermogenesis. Other important actions of thyroid hormones include a generalized increase in protein turnover (i.e. breakdown *and* synthesis), an increase in cardiac output caused by the enhancement of the effects of adrenaline (epinephrine) at β-adrenoceptors (Chapters 7 and 49), and a strong lipolytic effect that arises from the potentiation of responses to cortisol, glucagon, growth hormone and adrenaline. These actions can be described as generally catabolic (Chapter 43), but it should be noted that low doses of thyroid hormones have an overall anabolic action and that the hormones are essential to normal postnatal growth.

Disorders of the thyroid gland

Lack of dietary iodide or a failure of iodide uptake mechanisms in the thyroid gland produces the conditions of **hypothyroidism**. In fetal and neonatal life, underproduction of thyroid hormones causes inadequate somatic and neural development and gives rise to **cretinism**, a condition characterized by subnormal stature and mental ability. In adults, the main symptoms of thyroid insufficiency are lethargy, sluggishness and an intolerance to cold. In severe cases, there is excess production of water-retaining mucoproteins in subcutaneous tissues, giving rise to tissue bloating, known as **myxoedema**. Such conditions are treated with injections of T_4. When the cause of hypothyroidism is an insufficiency of iodide intake, cells of the thyroid gland undergo hypertrophy and the gland becomes enlarged to form a **goitre**. This (now very uncommon) condition is treated by ensuring an adequate supply of dietary iodide. The overproduction of T_3 and T_4 leads to **hyperthyroidism** (Graves' disease), characterized by **exophthalmia** (bulging eyes), increased behavioural excitability, tremor, weight loss and chronic tachycardia (high heart rate). The last of these symptoms can eventually lead to ventricular arrhythmias and/or heart failure, and so treatment, usually surgical removal of part of the gland or antithyroid drugs, is highly recommended.

46 Growth factors

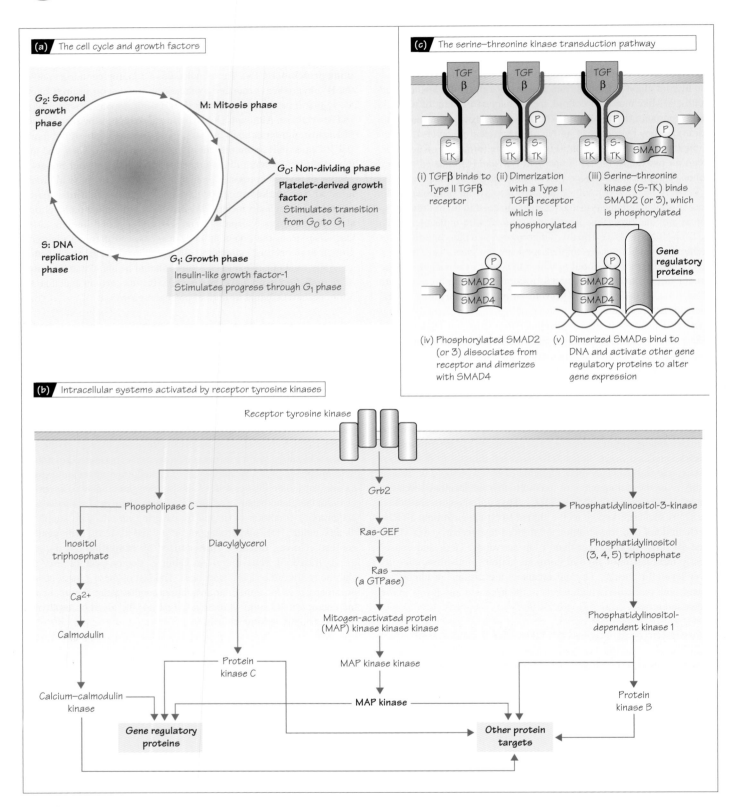

Physiology at a Glance, Third Edition. Jeremy P.T. Ward and Roger W.A. Linden.

 © 2013 John Wiley & Sons, Ltd. Published 2013 by John Wiley & Sons, Ltd.

For an embryo to develop into an adult, its cells must increase in number by the process of division (**mitosis**) and grow in size (**hypertrophy**). As they mature, cells develop specializations according to the tissue of which they are a part (**differentiation**). In tissue development, excess production of cells is the norm, so that the final shaping of organs depends on programmed death (**apoptosis**) of supernumerary cells. Some tissues, such as nerve and skeletal muscle, reach a stage of terminal differentiation in adulthood and undergo no further cell division. However, most cells in the adult retain the ability to divide, allowing tissues (e.g. blood vessels, bones) to remodel or repair themselves as required. Some cells face particularly high rates of attrition (e.g. enterocytes in the gut lining, skin cells, hair follicles) and are produced continuously throughout life. The processes of mitosis, cell growth and apoptosis are controlled by a large number of systemic and local peptide hormones, known as **growth factors** (Chapter 42). To varying extents, these factors stimulate mitosis (they are **mitogens**), promote growth (a **trophic** effect) and inhibit apoptosis (promote cell **survival**).

Growth factor families and their receptors

Growth factors are classified into a number of families based on common amino acid sequences and the types of receptor that they activate. **Neurotrophins**, which include **nerve growth factor (NGF)**, are important chemical signals in the development of the nervous system and are potent survival factors for neurones in adults. The **epidermal growth factor (EGF)** family includes EGF itself and transforming growth factor-α (TGFα), both of which are mitogens in a wide range of tissues, including the gut and skin. **Fibroblast growth factors (FGF-1–24)** are strongly mitogenic and induce the production of new blood vessels (**angiogenesis**). The **transforming growth factor-β (TGFβ)** superfamily includes a number of bone-transforming proteins (Chapter 43) and is crucial in embryogenesis and the development and remodelling of structural tissues. The origins of **platelet-derived growth factor (PDGF)** are self-explanatory. It stimulates division, growth and survival in a number of cell types, and is important in tissue repair after injury. **Insulin** and **insulin-like growth factors (IGF-1** and **IGF-2)** have similar structures but rather different actions: insulin promotes anabolic activity generally (Chapter 40), whereas the IGFs are mitogenic, trophic and act as survival factors for several cell types. Numerous other hormones have mitogenic properties, e.g. the stimulation of red blood cell production by **erythropoietin** (Chapter 8) and white cell production by **cytokines** (Chapter 10) means that these hormones are also described as growth factors.

Mitosis occurs during the **cell cycle** (Fig. 46a). Some mitogens, including PDGF, stimulate transition from the non-dividing state (G_0) into the growth phase of the cycle (G_1), whereas others, such as EGF and IGF-1, stimulate progress through G_1. With the exception of TGFα, erythropoietin and the cytokines, growth factors work by activating receptor tyrosine kinases (Chapter 43; Fig. 46b). Binding of the hormone leads to phosphorylation of the tyrosine residues of a number of important intracellular proteins, including phospholipase C, Grb2 and phosphatidylinositol-3 kinase, eventually leading to the production of more kinases: **protein kinases C and B**, **calcium-calmodulin kinase (CAM kinase)** and **mitogen-activated protein kinase (MAP kinase)** (Fig. 46b). These enzymes have many targets within the cell, but MAP kinase, in particular, enters the nucleus and activates immediate to early genes, such as *c-fos* and *c-jun*. The products of these genes are transcription factors, driving the expression of further genes, such as those that produce G_1 **cyclins**, proteins that are required for cell division. The MAP kinase pathway appears to be the main intracellular signalling system for the stimulation of mitosis. The TGFβ family exerts its effects through **receptor serine–threonine kinases** that phosphorylate their target proteins at serine and threonine residues. The pathway activated by these receptors involves proteins called **SMADs** [the name is derived from genes that code for similar proteins in *Drosophila melanogaster* (fruit fly) and *Caenorhabditis elegans*, a nematode worm]. SMAD-2 and/or SMAD-3 is phosphorylated while it is attached to the receptor; it then dissociates to dimerize with SMAD-4, forming a complex that directly activates gene regulatory proteins (Fig. 46c). Growth hormone, erythropoietin and the cytokines activate receptors that signal through **Janus kinases (JAKs**; Chapter 47).

Growth factors and cancer

Cell division and growth are strictly controlled so that organs do not invade the space needed for other tissues. When this process is deranged, cancers are formed. Cancer cells do not recognize the normal constraints of organ growth or the limits to the number of divisions to which cells are normally subjected, and are unusually mobile. These features make cancer cells extremely dangerous, as they supplant healthy tissues and cause fatal damage to physiological systems. Cancerous growths start with mutations in particular genes (**oncogenes**) that impact on cell division and/or apoptosis. *Ras* genes, which produce the Ras GTPases that are key mediators in the MAP kinase pathway (Fig. 46b), are commonly found to be defective in human tumours. In view of the importance of this pathway in mitogenesis, it is not difficult to see how the abnormal activation of these genes could lead to excessive cellular proliferation. In this situation, the signals involved in normal tissue growth provide the driving force for tumour growth and survival. EGF, in particular, has been associated with the maintenance of colorectal and breast cancers, and anti-EGF drugs are showing some promise as tumour-controlling agents.

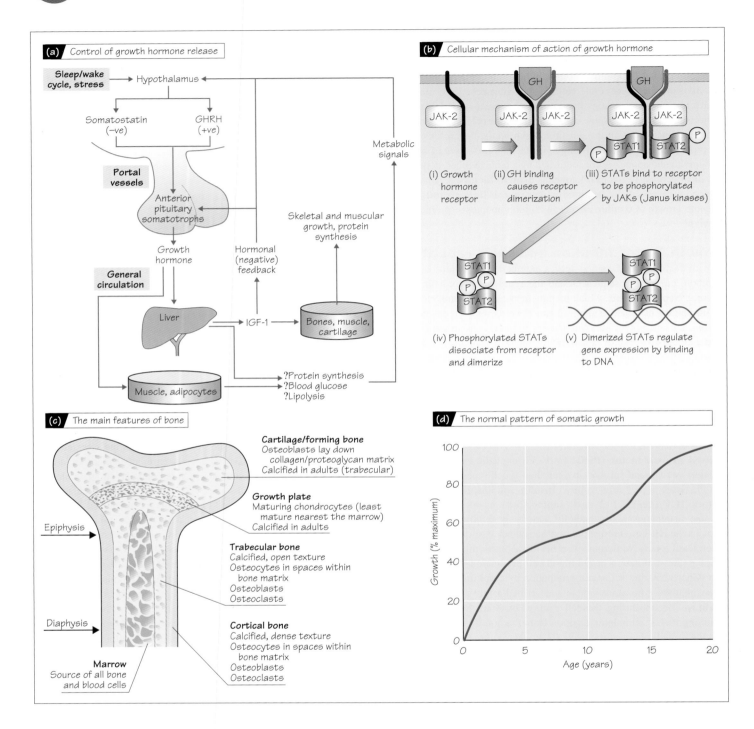

(a) Control of growth hormone release

Sleep/wake cycle, stress → Hypothalamus

Somatostatin (−ve) ← → GHRH (+ve)

Portal vessels

Anterior pituitary somatotrophs

Growth hormone

General circulation

Liver → IGF-1 → Bones, muscle, cartilage

Hormonal (negative) feedback

Skeletal and muscular growth, protein synthesis

Metabolic signals

Muscle, adipocytes → ?Protein synthesis, ?Blood glucose, ?Lipolysis

(b) Cellular mechanism of action of growth hormone

(i) Growth hormone receptor

(ii) GH binding causes receptor dimerization

(iii) STATs bind to receptor to be phosphorylated by JAKs (Janus kinases)

(iv) Phosphorylated STATs dissociate from receptor and dimerize

(v) Dimerized STATs regulate gene expression by binding to DNA

(c) The main features of bone

Epiphysis

Diaphysis

Marrow
Source of all bone and blood cells

Cartilage/forming bone
Osteoblasts lay down collagen/proteoglycan matrix
Calcified in adults (trabecular)

Growth plate
Maturing chondrocytes (least mature nearest the marrow)
Calcified in adults

Trabecular bone
Calcified, open texture
Osteocytes in spaces within bone matrix
Osteoblasts
Osteoclasts

Cortical bone
Calcified, dense texture
Osteocytes in spaces within bone matrix
Osteoblasts
Osteoclasts

(d) The normal pattern of somatic growth

Growth (% maximum) vs Age (years)

Physiology at a Glance, Third Edition. Jeremy P.T. Ward and Roger W.A. Linden.
 © 2013 John Wiley & Sons, Ltd. Published 2013 by John Wiley & Sons, Ltd.

Growth and development are entirely under endocrine control. The key signals involved in these processes are **growth hormone**, thyroid hormones (Chapter 45), sex steroids (Chapters 50 and 51) and growth factors (Chapter 46). Normal growth depends on the interplay between all of these factors. In development, there are two periods of particularly rapid growth: during pregnancy and up to the 2 years immediately after birth, and around the time of puberty (Fig. 47d).

Growth hormone

Growth hormone (GH; also known as **somatotrophin**) provides the main drive for growth. It is a protein released from pituitary somatotrophs under hypothalamic control (Fig. 47a) that stimulates growth in muscles, bones and connective tissue. It is essential for normal growth both before and after birth. The release of the hormone increases immediately after birth before subsiding to a low level for most of prepubertal life. There is another surge in release around the time of puberty, after which plasma concentrations again fall and then continue to decline steadily into old age. The release of the hormone varies throughout the day, with the highest levels achieved during deep sleep. The episodic appearance of growth hormone in the blood is driven by hypothalamic **growth hormone-releasing hormone** (**GHRH**), and **somatostatin** (**SST**), which inhibits growth hormone release (Chapter 44; Fig. 47a). The growth hormone receptor is linked to an intracellular enzyme, **Janus kinase-2** (**JAK-2**) (Fig. 47b). Once activated, this enzyme binds and phosphorylates **signal transduction and activation of transcription** (**STAT**) proteins, which consequently modify gene transcription. To provide energy for growing tissues, growth hormone has an anti-insulin action in increasing plasma glucose and stimulating lipolysis (Chapter 43). However, its overall effect is anabolic, increasing protein synthesis in many tissues. Most of its effects on growth arise from the stimulation of the release of insulin-like growth factor-1 (IGF-1) (Chapter 46) into the circulation, mainly from the liver. The lifetime release of growth hormone is regulated by the genetic factors that determine body size, but full expression of its effects requires adequate supplies of metabolic fuels and the presence of the other hormones mentioned above. In the short term, it is also liberated in response to stress and exercise.

The overproduction of growth hormone in children is associated with **gigantism**, and underproduction with **dwarfism**, which is much more common. Dwarfism is currently treated with human growth hormone manufactured by genetically engineered bacteria. Growth retardation can also result from defects in the GH receptor, or problems with IGF-1 production or action. Excess growth hormone release in adults leads to disproportionate growth of the bones of the face and limb extremities, a condition known as **acromegaly**.

Bone growth and remodelling

The bones are a major target for growth hormone. They are composed of an organic matrix made up of the structural protein **collagen**, combined with glycoproteins, that forms a framework within which the mineral **hydroxyapatite** $[Ca_{10}(PO_4)_6(OH)_2]$ is deposited. There are two main varieties of bone structure. **Cortical** or compact bone has a dense structure and provides most of the strength of the skeleton. It forms the outer layer of all bones and is particularly prevalent in the **diaphyses** (shafts) of limb bones. **Trabecular** or spongy bone has a more open structure than cortical bone and surrounds the marrow. Axial bones, such as the vertebrae, and the ends (**epiphyses**) of long bones are largely composed of trabecular matrix (Fig. 47c). In development, bones grow from the interface between the epiphysis and the diaphysis (the **growth plate**). The elongation of bones involves the laying down of new collagen matrix at the growth plate by rapidly dividing **chondrocytes**, followed by **calcification** (hydroxyapatite deposition) through the action of **osteoblasts**. When growth is complete at about 20 years of age (Fig. 47d), the growth plate itself becomes calcified and bone elongation ceases. This stage is known as **epiphyseal closure**, a process driven by the high levels of sex steroids present at puberty. Even in adults, bones remain dynamic structures, with substantial proportions of the skeleton (25% of trabecular bone and 3% of cortical bone) being replaced by new growth every year. Osteoblasts develop into **osteocytes**, cells with numerous processes that settle into spaces in the bone matrix. Osteocytes maintain the integrity of the matrix, but can also secrete acids that dissolve hydroxyapatite and thus provide free Ca^{2+} to the circulation when required (Chapter 48). **Osteoclasts** are large cells similar to macrophages (Chapter 10) that remove old bone matrix so that it can be replaced by new material. Osteoblasts, osteocytes and osteoclasts are all present in mature bone. The collective activity of these cells allows bone to be remodelled throughout life to cope with changes in skeletal stresses, and plays an essential role in the repair of broken bones. All bone cells differentiate from bone marrow stem cells. Systemic IGF-1 and locally produced IGF-1 and IGF-2 (Chapter 46) stimulate the division, differentiation and matrix-secreting activity of osteoblasts and chondrocytes (which are also involved in cartilage formation), whereas members of the transforming growth factor-β (TGFβ) family of growth factors are thought to provide the same stimuli for osteoclasts.

Osteoporosis

After the menopause women lose bone mass, leading to a weakening of the skeleton with a consequent increase in the likelihood of fractures in older women. This is due to the reduced secretion of sex steroids from the ovaries (Chapter 50), which normally suppress the production of the cytokine **interleukin-6** (**IL-6**) in bones. High levels of IL-6 stimulate the differentiation of osteoclasts, so that bone resorption outstrips the laying down of new matrix and more bone is removed than is replaced. The condition can be successfully treated by the administration of oestrogen (**hormone replacement therapy**). Recent evidence suggests that bone destruction in **rheumatoid arthritis** may also be driven by cytokines.

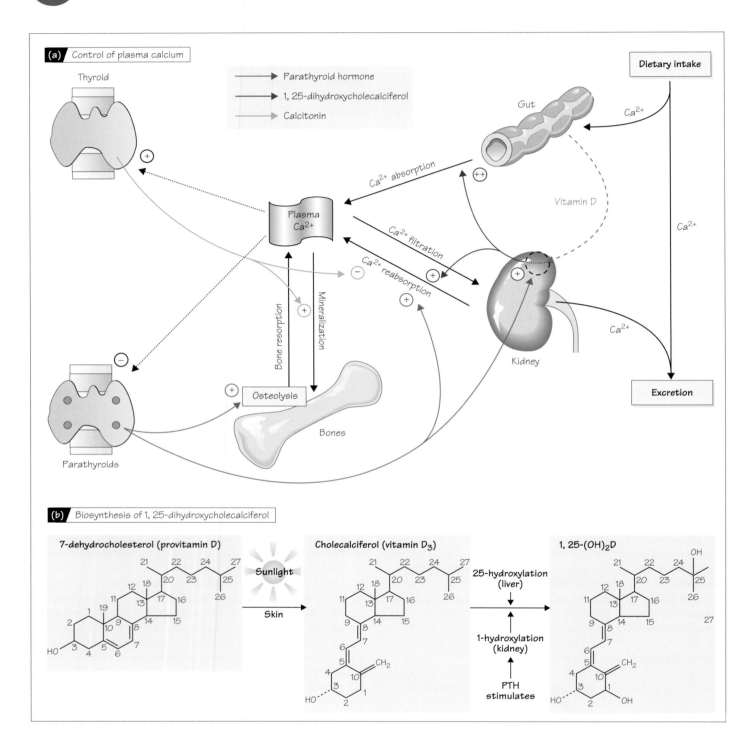

(a) Control of plasma calcium

Thyroid

→ Parathyroid hormone
→ 1, 25-dihydroxycholecalciferol
→ Calcitonin

Dietary intake

Gut

Ca^{2+}

Ca^{2+} absorption

Plasma Ca^{2+}

Vitamin D

Ca^{2+}

Ca^{2+} filtration

Ca^{2+} reabsorption

Bone resorption

Mineralization

Osteolysis

Kidney

Ca^{2+}

Parathyroids

Bones

Excretion

(b) Biosynthesis of 1, 25-dihydroxycholecalciferol

7-dehydrocholesterol (provitamin D)

Sunlight

Skin

Cholecalciferol (vitamin D₃)

25-hydroxylation (liver)

1-hydroxylation (kidney)

PTH stimulates

1, 25-(OH)₂D

Physiology at a Glance, Third Edition. Jeremy P.T. Ward and Roger W.A. Linden.

 © 2013 John Wiley & Sons, Ltd. Published 2013 by John Wiley & Sons, Ltd.

In cells, Ca^{2+} ions are used to trigger many key physiological events, including muscle contraction (Chapter 12), the release of neurotransmitters (Chapter 7), the release of hormones (Chapters 43 and 44), secretion from exocrine glands and the activation of many important intracellular enzymes, such as the calcium-calmodulin kinases (CAM kinases) (Chapter 46) and nitric oxide synthase (Chapters 21 and 51). Ca^{2+} ions are triggers in so many crucial events that free intracellular Ca^{2+} must be maintained at a very low level (Chapter 2). Most is stored in the endoplasmic reticulum or mitochondria. External to cells, Ca^{2+} ions contribute to the blood clotting cascade (Chapter 9) and the normal functioning of Na^+ ion channels (Chapter 4). When extracellular Ca^{2+} is too low, Na^+ channels open spontaneously, leading to involuntary contractions of skeletal muscles, described as **hypocalcaemic tetany**. This is the clinical sign of low plasma Ca^{2+}. It is evident that Ca^{2+} levels in plasma must be very carefully controlled, a function performed by the coordinated activity of three hormones: **parathyroid hormone (PTH)**, **1,25-dihydroxycholecalciferol [1,25-(OH)₂D]** and **calcitonin** (Fig. 48a).

Parathyroid hormone and calcitonin

PTH, a peptide of 84 amino acids, is the major controller of free calcium in the body. It is released from **chief cells** of the four (or more) **parathyroid glands** located immediately behind the thyroid gland, when the plasma concentration of Ca^{2+} decreases. The ion is detected by a membrane-bound receptor protein expressed by chief cells. When Ca^{2+} ions bind to the receptor, intracellular levels of cyclic adenosine monophosphate (cAMP) (Chapter 3) decrease and the release of PTH is *inhibited*. PTH increases the plasma levels of Ca^{2+} by activating specific membrane receptors in bone, gut and kidney. In bone, the immediate effect of PTH is to stimulate **osteocytic osteolysis** of bone crystals to release Ca^{2+} ions (Chapter 47). After a longer time, PTH also increases osteoclast activity (Chapter 47) to gain access to more of the bone mineral. In the gut, PTH, acting in concert with 1,25-dihydrocholecalciferol, enhances the absorption of Ca^{2+} ions. In the kidney, the same combination of hormones enhances the reabsorption of Ca^{2+} from the renal tubules and simultaneously decreases the reabsorption of PO_4^{3-} ions (Chapter 34). PTH also stimulates the kidney to produce more 1,25-dihydrocholecalciferol. Thus, PTH leads the response to a fall in plasma Ca^{2+} by releasing ions stored in bone, conserving ions filtered by the kidney and enhancing the intake of new ions from the gut (Fig. 48a). The effects of PTH are in each case mediated by stimulating an increase in cAMP in the target cells. **Calcitonin** is a 32-amino acid peptide released from C cells of the thyroid gland in response to high levels of plasma Ca^{2+} ions. C cells carry the same Ca^{2+} receptor as parathyroid chief cells. Calcitonin inhibits bone resorption by osteocytes and may inhibit reabsorption in the kidney, so reducing plasma

levels of the ion. The fact that complete removal of the thyroid gland causes no obvious problems with calcium homeostasis has led some physiologists to doubt the significance of this hormone in the normal control of Ca^{2+} ions.

Vitamin D and 1,25-dihydroxycholecalciferol

Vitamin D is an umbrella term for two molecules: **ergocalciferol** (vitamin D₂) and **cholecalciferol** (vitamin D₃). Both are derivatives of provitamin D (dehydrocholesterol; Fig. 48b) and the primary source of supply is the diet, with ergocalciferol derived from plants and yeasts and cholecalciferol from animal (particularly dairy) products. Unusually for a vitamin, cholecalciferol can be manufactured within the body via a reaction that is enabled by ultraviolet irradiation of the skin. The D vitamins are then converted to 1,25-dihydroxycholecalciferol in the kidney (Fig. 48b). The final reaction is the slowest step in the process and therefore regulates the speed of the entire chain of reactions (i.e. it is rate limiting); it is under the influence of PTH. 1,25-Dihydroxycholecalciferol has a steroid-like structure and is sometimes referred to as a **sterol**. Its receptors are members of the superfamily of steroid receptors and are located *inside* target cells. The hormone–receptor complex binds to response elements on deoxyribonucleic acid (DNA) to drive the transcription of genes. **Calcium-binding protein**, which is thought to promote calcium transport across epithelia (Chapter 34), is the product of one of the genes activated by 1,25-dihydroxycholecalciferol. The major action of 1,25-dihydroxycholecalciferol is to enable Ca^{2+} absorption from the gut. Without the hormone, Ca^{2+} uptake is severely impaired to the point at which intake of the hormone is insufficient to maintain body stores. This leads to the increased release of PTH and resorption of bone. A lack of D vitamins in children leads to inadequate calcification of bones, which become malformed. This leads to the characteristically bowed limbs seen in **rickets**. This condition was common in the early part of the 20th century, but was virtually eliminated in the UK by the introduction of free school milk. Insufficiency of vitamin D in adults leads to bone wasting, a condition known as **osteomalacia**, with symptoms similar to those of osteoporosis (Chapter 47). 1,25-Dihydroxycholecalciferol also promotes the reabsorption of Ca^{2+} from the kidney tubules. The effects of this hormone are generally augmented in the presence of PTH.

Other hormones affecting calcium

Growth-promoting hormones (growth hormone, thyroid hormones and sex steroids) tend to promote the incorporation of calcium into bones (Chapter 47). Excess corticosteroids (Chapter 49) inhibit calcium uptake from the gut and reabsorption from the kidney.

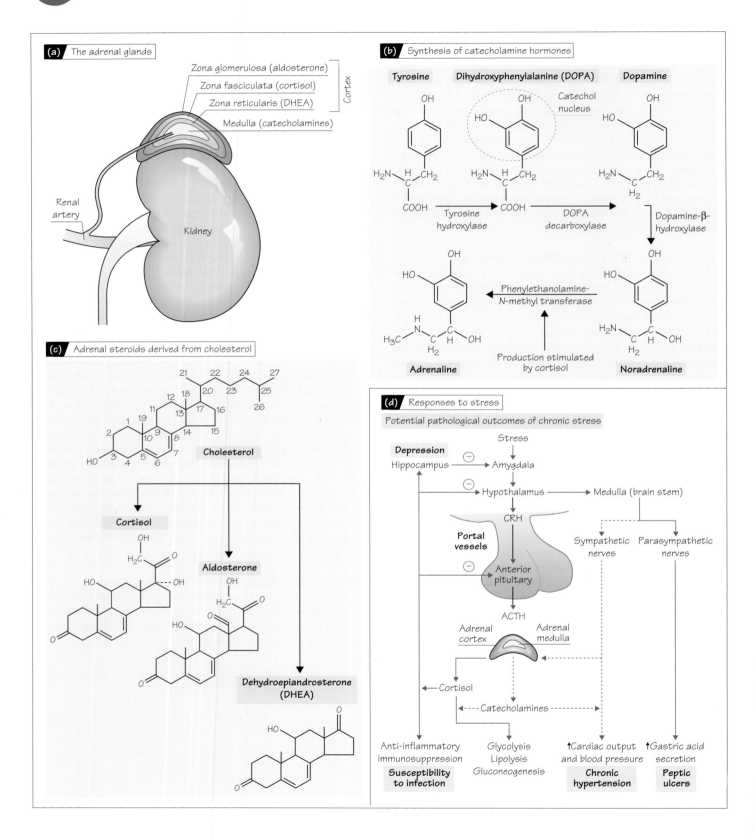

Physiology at a Glance, Third Edition. Jeremy P.T. Ward and Roger W.A. Linden.

108 © 2013 John Wiley & Sons, Ltd. Published 2013 by John Wiley & Sons, Ltd.

The adrenal glands are located just above each kidney (hence the name; Fig. 49a) and consist of two endocrine tissues of distinct developmental origins. The inner core (the **adrenal medulla**) releases the catecholamine hormones **adrenaline** (epinephrine) and **noradrenaline** (norepinephrine). It develops from neuronal tissue and is functionally part of the **sympathetic nervous system** (Chapter 7). The outer layers of the gland (the **adrenal cortex**) originate from mesodermal tissue and secrete steroid hormones, primarily under the control of the anterior pituitary gland (Chapter 44). Removal of the adrenal glands in animals results in death within a few days, which is thought to result from the loss of the ability to cope with **stress**.

The adrenal medulla

The **chromaffin cells** of the adrenal medulla manufacture and secrete **noradrenaline** (20%) and **adrenaline** (80%). These catecholamine hormones are derived from tyrosine by a series of steps catalysed by specific enzymes (Fig. 49b). The production of the rate-limiting enzyme, **phenylethanolamine-N-methyl transferase**, is stimulated by **cortisol**, providing a direct link between the functioning of the medulla and cortex. The secretion of catecholamines is stimulated by sympathetic preganglionic neurones located in the spinal cord (Chapter 7), so that the adrenal medulla functions in concert with the sympathetic nervous system, of which noradrenaline is the main neurotransmitter. Catecholamine release contributes to normal physiological functions, but is enhanced by stress (see below). Adrenaline and noradrenaline act through guanosine triphosphate-binding protein (G-protein)-coupled **adrenoceptors**. These are classified as α_1, α_2 and β_1–β_3. The hormones have the same effects in tissues as the stimulation of sympathetic nerves, with important stress-related responses being vasoconstriction (α_1), increased cardiac output (β_1) and increased glycolysis and lipolysis (β_2, β_3). These actions support increased physical activity. Noradrenaline has equal potency at all adrenoceptors, but adrenaline, at normal plasma concentrations, will only activate β-receptors (NB: higher levels do stimulate α-receptors). **Phaeochromocytoma** is a tumour of the adrenal medulla that leads to the excess production of catecholamines, with high blood pressure as the most immediately threatening symptom. It is treated by α-adrenoceptor antagonists and/or surgery.

The adrenal cortex

The cortex is made up of three zones of tissue: the outer **zona glomerulosa**, which releases **aldosterone**; the **zona fasciculata**, which produces **cortisol** and several related but less important hormones; and the inner **zona reticularis**, which secretes the androgen **dehydroepiandrosterone (DHEA)**. All of these secretions are steroids (Fig. 49c). Aldosterone is referred to as a **mineralocorticoid** as it controls the reabsorption of Na^+ and K^+ ions in the kidney (Chapter 35), whereas DHEA and its metabolite, **androstenedione**, provide an important source of androgens for females, contributing to hair growth and libido (Chapter 50). Cortisol and its analogues (such as cortisone) have powerful effects on glucose metabolism and are collectively classified as **glucocorticoids**, although they do have some mineralocorticoid

actions. The release of cortisol and DHEA is stimulated by **adrenocorticotrophic hormone (ACTH)** liberated from the pituitary gland (Chapter 44; Fig. 49d), whereas the secretion of aldosterone is stimulated by angiotensin II (Chapter 35). The effects of cortisol are mediated by intracellular receptors that translocate to the cell nucleus after binding the hormone. The cortisol–receptor complex binds to **glucocorticoid response elements** on deoxyribonucleic acid (DNA) to initiate gene transcription.

Cortisol is released during the course of normal physiological activity. The pattern of secretion is pulsatile, driven by activity in corticotrophin-releasing hormone (CRH) neurones of the hypothalamus (Chapter 44). There is usually a surge in cortisol release in the hour after waking. The primary stimulus for the increased release of glucocorticoids is **stress**, which is the result of exposure to adverse situations. The **stress response** is driven by the **amygdala**, part of the forebrain that stimulates: (i) activity in hypothalamic CRH neurones; (ii) activity in the sympathetic nervous system; (iii) activity in the parasympathetic nerves that cause acid secretion in the stomach (Chapter 38); and (iv) the feeling of fear (Fig. 49d). The stress response evolved to cope with immediate threats, such as predators, to which the appropriate physiological reaction is to prepare for physical activity. The actions of the two parts of the adrenal gland are complementary in this respect. Catecholamines are released from the medulla to produce a rapid increase in cardiac output and the mobilization of metabolic fuels. Corticosteroids produce a slower, more sustained response, increasing the amount of glucose in the plasma (Chapter 43) by: (i) increasing glycolysis and gluconeogenesis in the liver (Chapter 40); (ii) reducing glucose transport into storage tissues; (iii) increasing protein catabolism with a consequent release of amino acids from all tissues other than the liver; and (iv) increasing the mobilization of lipids from adipose tissue. High levels of glucocorticoids also suppress the activity of immune cells to produce an anti-inflammatory effect, and can mimic the actions of aldosterone on the kidney to retain Na^+ and lose K^+ ions. The stress response is appropriate as long as the stress is relieved promptly. Unfortunately, modern life places many of us in positions in which stress is prolonged. This can lead to chronic hypertension, gastric ulceration, immunosuppression and depression (Fig. 49d). Glucocorticoid derivatives, such as **dexamethasone**, are widely used as anti-inflammatory agents in conditions such as arthritis and asthma. Chronically high levels of glucocorticoids eventually cause weakening of the skin, muscle wasting, reduction in bone strength, increased rates of infection due to immunosuppression, and can damage nerve cells in the **hippocampus** that are part of a feedback circuit controlling responses to stress (Fig. 49d). Thus, the long-term therapeutic use of steroids must be very carefully monitored, especially in the young where normal growth may be affected. Diseases of the adrenal cortex include **Cushing's syndrome**, which results from the excessive release of glucocorticoids and has a range of symptoms similar to those described above, and **Addison's disease**, which is the result of adrenocortical hypoactivity and is characterized by symptoms of hypoglycaemia, weight loss and skin pigmentation.

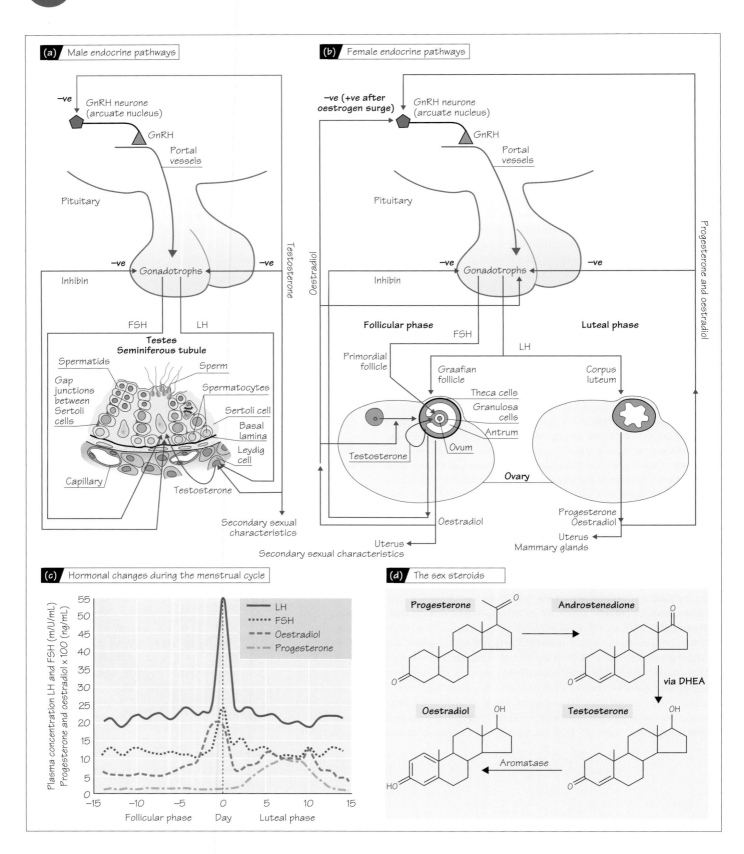

Physiology at a Glance, Third Edition. Jeremy P.T. Ward and Roger W.A. Linden.

 © 2013 John Wiley & Sons, Ltd. Published 2013 by John Wiley & Sons, Ltd.

Reproductive function in males and females is controlled by common hormonal systems based on the hypothalamic control of the pituitary **gonadotrophins**, individually known as **luteinizing hormone (LH)** and **follicle-stimulating hormone (FSH)**. These glycoproteins are released from the **gonadotrophs** of the anterior pituitary gland under the influence of **gonadotrophin-releasing hormone (GnRH**; Chapter 44) (Fig. 50a,b). Failure of GnRH release is one cause of infertility. It is released in pulses at intervals of 1–3 h in both males and females, a pattern that is accurately reflected in plasma levels of LH. The pulsatile pattern of GnRH secretion is essential for normal reproductive activity, as continuous exposure of gonadotrophs to the hormone leads to a rapid desensitization of the gonadotrophs and a reduction in the release of gonadotrophins. The releasing hormone acts through receptors coupled to G_q (Chapter 3) to stimulate the release and manufacture of the gonadotrophins.

Actions of gonadotrophins

The gonadotrophins produce their effects via interactions with guanosine triphosphate-binding protein (G-protein)-coupled receptors that activate the intracellular production of cyclic adenosine monophosphate (cAMP) (Chapter 3). In the male, LH acts on the **Leydig cells** of the testes to stimulate the production of the steroid **testosterone**, which acts in concert with FSH on **Sertoli cells** of the **seminiferous tubules** to support **spermatogenesis** (Fig. 50a). Sperm are generated in a two-stage meiosis from spermatocytes via spermatids. Spermatogenesis proceeds most efficiently at a temperature of 34 °C, which is why the testes are located outside the body cavity. A normal adult male produces some 2×10^8 sperm per day, a process that carries on from puberty until the end of life. Sertoli cells also produce **inhibin**, a peptide feedback signal that specifically inhibits the release of FSH from the anterior pituitary.

The situation in females varies over time according to the **menstrual cycle** (Fig. 50b,c), which lasts for around 28 days but is also ultimately driven by the activity of the hypothalamic GnRH neurones. After puberty, the ovaries contain about 400 000 **primordial follicles**, each of which contains an **ovum** (or **oocyte**) in an arrested state of meiosis. All follicles are present at birth and no new gametes are formed after this time. Small groups of follicles begin to mature spontaneously throughout reproductive life, but only those for which development coincides with the appropriate phase of the cycle reach the stage of ovulation. In the first part of the cycle (the **follicular phase**), LH acts on **theca interna cells** in developing follicles to stimulate the production of testosterone, which is converted to **oestrogens** (mainly **oestradiol**; Fig. 50b) by **aromatase** enzymes in follicular **granulosa cells** under the influence of FSH. Granulosa cells also produce inhibin, which suppresses FSH release. In the follicular phase, oestrogens promote the growth of the uterine endometrial lining and the release of watery secretions at the cervix that enhance the transit of sperm into the uterus. Oestrogens also stimulate the production of

LH receptors in granulosa cells. During this time, the actions of FSH and oestrogens stimulate maturing follicles within the ovary, only the largest of which will normally undergo **ovulation**. The remainder wither away by the process of **atresia**. Ovulation occurs at about day 14 of the cycle (Fig. 50c). It is initiated by a large increase in the release of oestradiol from the granulosa cells, stimulated by their newly developed LH receptors. Normally, oestrogens act as a negative feedback signal, inhibiting LH release (Fig. 50b), but the large amounts secreted by the mature follicle *stimulate* LH release, i.e. the system switches from negative to *positive* feedback. This leads to a massive increase in the release of LH, which causes the wall of the most developed follicle to rupture and releases the ovum into the nearest **oviduct** to await fertilization (Chapter 52). Following ovulation, the granulosa cells undergo **hypertrophy** (growth) and the ruptured follicle develops into the **corpus luteum**, and the cycle enters the **luteal phase**. The corpus luteum produces **progesterone** (Fig. 50b), as well as oestrogens, in response to stimulation by LH. Progesterone prepares the reproductive tract for pregnancy, stimulating further growth of the uterine endometrium and altering the nature of cervical secretions to discourage the entry of sperm into the uterus. If fertilization does not occur, the corpus luteum undergoes **luteolysis** after roughly 14 days, a process that results from the reduced ability of LH to support the corpus luteum. In the absence of progesterone and oestradiol, the endometrial lining degenerates and is shed in the process of **menstruation**, followed by the onset of a new cycle. After 30–40 years of menstrual activity, the exhaustion of ovarian follicles causes the female system to enter the **menopause**, after which reproduction is no longer possible. Circulating levels of sex steroids are greatly reduced, leading to drying of the secretory glands in the reproductive tract and other symptoms, including circulatory changes that cause hot flushes. The most pernicious outcome of the menopause is **osteoporosis** (Chapter 48).

All sex steroids exert their effects by interacting with intracellular receptors that bind to deoxyribonucleic acid (DNA) response elements, and thus induce changes in gene expression. Some of the actions of testosterone are actually mediated by its conversion to the more active **dihydrotestosterone**, produced within the target cells by the action of the enzyme **5-α-reductase**.

Hormonal contraceptives

Human fertility control currently rests firmly on the use by women of hormonal contraceptives. These agents can contain a mixture of synthetic oestrogens and progestogens (analogues of progesterone), or progestogens only, and are administered as daily tablets, depot injections that last for several months, or as long-term (5 years) uterine implants. They probably have multiple sites of action, affecting negative feedback signals to suppress gonadotrophins, the consistency of cervical mucus to prevent sperm penetration, and the sensitivity of the uterine lining to prevent implantation of the embryo.

LIVERPOOL JOHN MOORES UNIVERSITY
LEARNING SERVICES

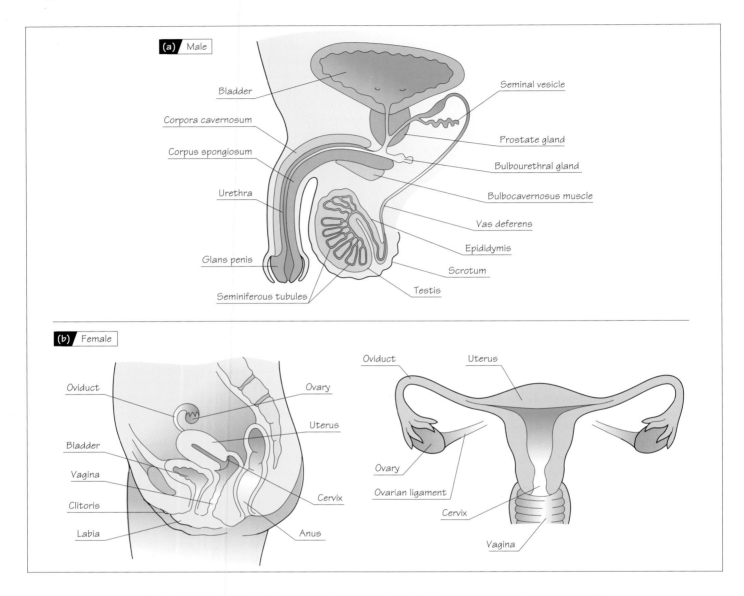

Table 51 Primary and secondary sexual characteristics that develop during puberty

Females (mostly stimulated by oestrogens)	Males [stimulated by (dihydro)testosterone]
Growth and maturation of ovaries	Growth and maturation of testes, including descent into scrotum
Growth of external genitalia	Growth of external genitalia and pubic hair
Growth of breasts	Increased size of larynx, leading to deeper voice
Keratinization of vaginal mucosa, enlargement of uterus	Increased muscle mass and strength
Deposition of fat around hips and thighs	Thickened skin
Growth of pubic hair (*stimulated by adrenal androgens*)	Increased and thickened body hair
	Increased bone mass (*also requires oestrogen*)

Physiology at a Glance, Third Edition. Jeremy P.T. Ward and Roger W.A. Linden.
 © 2013 John Wiley & Sons, Ltd. Published 2013 by John Wiley & Sons, Ltd.

Sexual differentiation

Gender is determined by the presence of X and Y chromosomes in the genome. Two X chromosomes provide the female genotype, whereas X and Y chromosomes together give a genetic male. Undifferentiated gonads are apparent after about 4–6 weeks of gestation, and both **Müllerian ducts**, which eventually form the uterus and Fallopian tubes, and **Wolffian ducts**, which form the vas deferens, epididymis and seminal vesicles, are present. The early gonads secrete steroids just as they do in the adult, and these hormones determine the sexual phenotype. In the absence of the *Sry* gene on the Y chromosome and thus testosterone, the Müllerian ducts continue to differentiate whilst the Wolffian ducts regress. The development of reproductive organs and brain connectivity therefore defaults to a female pattern which is dependent on the secretion of oestrogens.

The *Sry* gene is thought to be responsible for the establishment of testicular development and **Leydig cells** which secrete testosterone. Testosterone stimulates the development of the male genitalia (Fig. 51a) and the organization of neuronal systems in the brain that are involved in sexual function and behaviour. Notably, there is marked growth in the **sexually dimorphic nucleus** of the medial preoptic area of the hypothalamus and in the spinal nucleus that controls the **bulbocavernosus** muscle which is involved in ejaculation. Curiously, testosterone has to be converted to oestrogen by brain aromatases to have these effects. The fetal testis also secretes **anti-Müllerian hormone** (AMH) which causes regression of Müllerian ducts and thus prevents the uterus and Fallopian tube from developing.

Puberty

Although active before birth, the gonadotrophic axis quickly becomes quiescent after parturition and remains so until the onset of puberty at 8–14 years. The trigger for this remains obscure, but may result from endogenous activation of brain pattern generating circuits that stimulate **gonadotrophin-releasing hormone** (GnRH) neurones. **Body mass**, signalled via circulating levels of leptin (Chapter 43) and insulin-like growth factor-1 (IGF-1) (Chapter 46), are important permissive factors in females and undernutrition is associated with failure of the menstrual cycle. Puberty begins when GnRH stimulates cyclic release of **luteinizing hormone** (LH) and **follicle-stimulating hormone** (FSH) from the anterior pituitary (Chapter 50), first at night and then throughout the day. LH stimulates release of testosterone from Leydig cells in males and follicular oestrogens in females, and FSH the onset of spermatogenesis in males and follicle growth in females; they therefore act **synergistically**. This is accompanied by the many physical changes associated with the final growth into an adult (Table 51). The appearance of secondary sexual characteristics in the male is thought to be largely stimulated by the testosterone metabolite, dihydrotestosterone. In females, the onset of the cyclic release of LH and thus oestrogens gives rise to the beginning of menstruation (**menarche**) and the development of the mature female body pattern (Table 51). The end of puberty marks the onset of full sexual maturity and the conclusion of somatic growth (Chapter 47). Figure 51a,b shows the mature male and female reproductive tracts.

Sexual function

Sexual attraction and behaviour in humans are the highly complex result of physiological factors, combined with societal and other psychological influences. The overall level of **libido** (sexual motivation) is set by the hypothalamus under the influence of higher centres and the hormonal environment. In males, sexual arousal arises from physical stimulation of the genitalia (a spinal reflex) or from psychological stimuli (by pathways descending from the hypothalamus via the brain stem) that activate sacral parasympathetic nerves (Chapter 7). The penis becomes erect as the result of the dilation of blood vessels entering the **corpora cavernosum** (the main erectile tissue) and **corpus spongiosum** (Fig. 51a). The enhanced flow of blood into the cavernous spaces increases tissue pressure and restricts venous drainage, causing a further build-up of pressure to make the penis fully erect. The parasympathetic nerves cause vasodilatation by the release of acetylcholine, vasoactive intestinal peptide and, primarily, **nitric oxide** (**NO**; Chapter 21). NO increases the manufacture of **cyclic guanosine monophosphate** (**cGMP**) in blood vessel smooth muscle cells to cause them to relax. Sildenafil (Viagra) inhibits the breakdown of cGMP and thus enhances erectile function. The female sexual response sometimes involves erection of the clitoris, but the main manifestations are relaxation of the smooth muscles of the vagina and an increase in mucous secretions that act as a lubricant. Again, these actions are brought about by the activation of parasympathetic nerves. The combined effects of the male and female sexual responses facilitate entry of the penis into the vagina (**intromission**). Frictional forces stimulate mechanoreceptors in the glans penis and the clitoris that eventually lead to reflex activation of the sympathetic nerves that causes **orgasm**. In the male, this involves peristaltic contractions of the epididymis to pump sperm into the urethra, where they are mixed with the secretions of the **bulbourethral gland**, the **seminal vesicle** and the **prostate gland** to form semen. The secretions provide, respectively, lubrication, energy (in the form of the sugar **fructose**) and an alkaline barrier against the acid conditions normally prevalent in the vagina. They also include high levels of **prostaglandins**, the arachidonic acid-derived local hormones that stimulate the motility of sperm and of the female tract. Further peristaltic contractions of the urethra, in combination with the action of the bulbocavernosus muscle, emit the semen bolus into the upper end of the vagina (**ejaculation**). The female orgasm, which may involve the release of pituitary **oxytocin** elicited by mechanical stimulation of the cervix (Chapter 52), results in rhythmic contractions of the vaginal and uterine muscles that promote the flow of semen into the uterus. Sperm move by means of their own motility and by the beating of cilia on the walls of the uterus, but only a few hundred sperm of the millions released in a single ejaculate will complete the 6-h journey from the vagina to the oviducts.

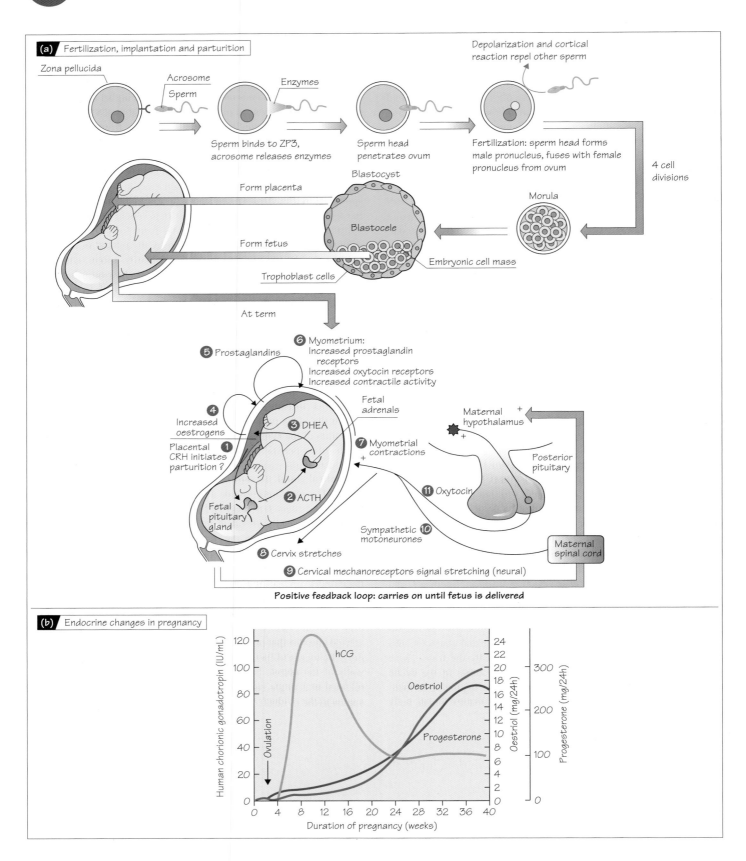

(a) Fertilization, implantation and parturition

Zona pellucida — Acrosome, Sperm

Sperm binds to ZP3, acrosome releases enzymes

Enzymes

Sperm head penetrates ovum

Depolarization and cortical reaction repel other sperm

Fertilization: sperm head forms male pronucleus, fuses with female pronucleus from ovum

4 cell divisions

Morula

Blastocyst

Form placenta

Blastocele

Form fetus

Trophoblast cells — Embryonic cell mass

At term

⑤ Prostaglandins

⑥ Myometrium: Increased prostaglandin receptors Increased oxytocin receptors Increased contractile activity

④ Increased oestrogens

③ DHEA

Fetal adrenals

Maternal hypothalamus +

+

Placental CRH initiates parturition ? ①

⑦ Myometrial contractions +

Posterior pituitary

② ACTH

Fetal pituitary gland

⑪ Oxytocin

Sympathetic motoneurones ⑩

Maternal spinal cord

⑧ Cervix stretches

⑨ Cervical mechanoreceptors signal stretching (neural)

Positive feedback loop: carries on until fetus is delivered

(b) Endocrine changes in pregnancy

Human chorionic gonadotropin (IU/mL) / Oestriol (mg/24h) / Progesterone (mg/24h)

hCG

Oestriol

Ovulation

Progesterone

Duration of pregnancy (weeks)

Physiology at a Glance, Third Edition. Jeremy P.T. Ward and Roger W.A. Linden.

 © 2013 John Wiley & Sons, Ltd. Published 2013 by John Wiley & Sons, Ltd.

Fertilization

The unfertilized ovum can survive for up to 24 h after ovulation, and sperm remain viable in the uterus for up to 5 days after ejaculation. The environment of the female tract triggers the **capacitation** of sperm. This is a prerequisite for fertilization that involves remodelling of the lipids and glycoproteins of the sperm plasma membrane, coupled with increased metabolism and motility. The ovum is surrounded by the **zona pellucida**, an acellular membrane bearing the glycoprotein **ZP3** that acts as a sperm receptor. Fertilization occurs in the oviduct, when a single capacitated sperm binds to ZP3 and undergoes the **acrosome reaction**. The acrosome is a body containing proteolytic enzymes that is attached to the sperm head (Fig. 52a). When a sperm binds to ZP3, the acrosomal enzymes are released to digest a pathway for the sperm to penetrate the ovum, within which the contents of the sperm head, including its genetic material, are deposited. This event leads to a chain of reactions that denies access to further sperm penetration. The ovum first undergoes electrical depolarization and then discharges granules that impair further sperm binding at the zona pellucida (the **cortical reaction**). In this way, fertilization is normally restricted to one sperm per ovum. Some 2–3 h after penetrating the ovum, the sperm head forms the **male pronucleus** which joins with the **female pronucleus** from the ovum (Fig. 52a). Fusion of the pronuclei combines the parental genetic material from the gametes to form the **zygote**.

Pregnancy

The zygote is propelled by cilia and muscular contractions of the Fallopian tube into the uterus, where it implants in the endometrium. During this journey, the zygote undergoes a number of cell divisions to form the **morula**, a solid ball of 16 cells that 'hatches' from the zona pellucida and develops into the **blastocyst**, in which embryonic cells are surrounded by **trophoblasts** (Fig. 52a). The trophoblasts are responsible for implantation, digesting away the uterine endometrial wall to form a space for the embryo, opening up a pathway to the maternal circulation (via the **spiral arteries** of the uterus) and forming the fetal portion of the **placenta**. The tissue engineering activities of trophoblasts are mediated by epidermal growth factor (EGF) (Chapter 46) and interleukin-1β. Implantation is complete within 7–10 days of fertilization, at which time the embryo and early placenta begin to secrete **human chorionic gonadotrophin (hCG)**. The appearance of hCG in the plasma and urine is one of the earliest signs of successful conception, and its detection forms the basis of pregnancy testing kits. hCG is a glycoprotein similar to LH that stimulates progesterone secretion from the corpus luteum. Progesterone levels rise steadily throughout pregnancy and fall sharply at term (Fig. 52b). This steroid ensures that the smooth muscle of the uterus remains quiescent during gestation (essential for a successful pregnancy), stimulates mammary gland development and prepares the maternal brain for motherhood. The placenta also secretes **chorionic somatomammotrophin**, a growth hormone-like protein that mobilizes metabolic fuels (Chapter 43) and promotes mammary gland growth, and oestrogen (mainly **oestriol**) that stimulates uterine expansion to accommodate the growing fetus. Fetal development occurs within a fluid-filled sac, known as the **amniotic membrane**, which provides a protective buffer against physical trauma. Pregnancy makes many physiological demands on the mother. The ventilation rate, cardiac output and plasma volume increase to supply fetal–maternal oxygen and water demands; the gastrointestinal absorption of minerals is enhanced; and the renal glomerular filtration rate (Chapter 32) rises to cope with fetal waste production.

Parturition

After some 40 weeks of gestation, the fetus is ready for life outside the uterus. The signal that initiates parturition in humans is still not fully understood, and there seems to be a difference between primates and other mammals. In primates, the primary signal is thought to arise from the **fetoplacental unit** (i.e. the fetus plus the placenta) as an increase in dehydroepiandrosterone (DHEA) production from the fetal adrenal cortex (Chapter 49), which may be driven by *placental* (rather than hypothalamic) production of corticotrophin-releasing hormone (CRH) (Chapters 44 and 49; Fig. 52a). DHEA is a precursor for oestrogen production by the placenta. As the placental aromatase enzymes are not rate limiting, an increase in DHEA, which is a precursor of oestrogen (Chapter 50), automatically increases oestrogen production. Whatever the initiating signal might be, the end result is an increase in the synthesis of prostaglandins E and F by fetal and uterine tissues, with concomitant increases in prostaglandin receptors in the uterine smooth muscle. The prostaglandins stimulate the production of uterine receptors for **oxytocin** and change the pattern of activity in the uterine myometrium from slow, gentle contractions to regular, deep contractions that eventually move the fetus into the cervix. The cervix, which is softened by prior release of the prostaglandins, dilates as the fetus is forced downwards. At this time, the amniotic membrane ruptures. Stretching of the cervix activates mechanoreceptors that stimulate a spinal sympathetic reflex which causes myometrial contraction and secretion of **oxytocin** from the posterior pituitary gland (Chapter 44; Fig. 52a). Oxytocin is a powerful stimulant of uterine smooth muscle that causes further contraction of the myometrium and pushes the fetus further into the cervix, resulting in further stimulation of mechanoreceptors and leading to the release of more oxytocin, i.e. this is a positive feedback system. The spinal reflex, aided by waves of oxytocin, generates large, regular contractions of the uterus that eventually expel the fetus and placenta through the vagina, completing the birth process. Oxytocin continues to be useful, as it limits maternal bleeding by causing vasoconstriction. In the fetus, oxytocin closes the **ductus arteriosus**, a blood vessel that shunts blood away from the pulmonary circulation *in utero*, but which would obviously hamper postpartum life should it remain open.

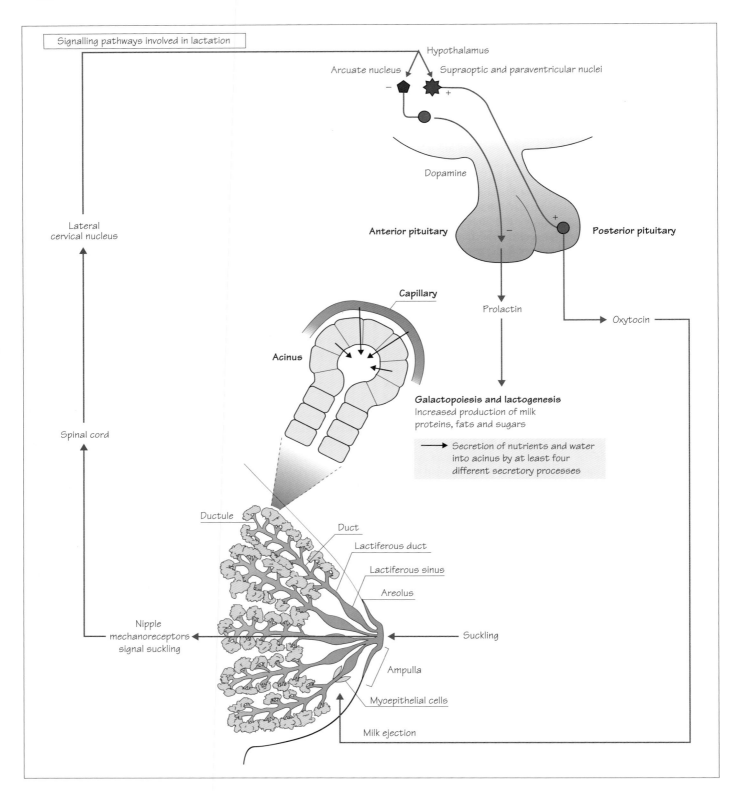

Signalling pathways involved in lactation

Hypothalamus

Arcuate nucleus   Supraoptic and paraventricular nuclei

Dopamine

Lateral
cervical nucleus

Anterior pituitary   Posterior pituitary

Capillary

Acinus

Prolactin   Oxytocin

Galactopoiesis and lactogenesis
Increased production of milk
proteins, fats and sugars

Secretion of nutrients and water
into acinus by at least four
different secretory processes

Spinal cord

Ductule
Duct
Lactiferous duct
Lactiferous sinus
Areolus

Nipple
mechanoreceptors
signal suckling   Suckling

Ampulla

Myoepithelial cells

Milk ejection

Physiology at a Glance, Third Edition. Jeremy P.T. Ward and Roger W.A. Linden.
 © 2013 John Wiley & Sons, Ltd. Published 2013 by John Wiley & Sons, Ltd.

Milk, which sustains mammalian infants through the first few months of life, is produced by the mammary glands (Fig. 53) under the influence of the pituitary protein hormone **prolactin** (Chapter 44). The glands comprise several lobules that are composed of **acini** (also called **alveoli**), similar in structure to the salivary glands and the exocrine pancreas (Chapters 37 and 40). The lobules empty into **lactiferous ducts**. As the ducts approach the **areola** (nipple), they open out to form **lactiferous sinuses** before narrowing again to emerge at the **ampulla** on the nipple. The ducts and sinuses are organized so that milk collects within them rather than flowing freely to the ampulla. They are lined by **myoepithelial cells** that contract to expel milk from the breast. Progesterone, oestrogen, prolactin, cortisol and growth hormone are all required to complete development of the mammary glands, which occurs during the late stages of pregnancy; for the rest of adult life the glandular tissue is rather small. Milk is formed by intense activity of the epithelial cells lining the acinus. The acinar secrete fats (triglycerides), proteins (principally **casein**, α-lactalbumin and lactoglobulin B) and sugars (mostly **lactose**) to produce an isotonic liquid that is roughly 4% fat, 1% protein and 7% sugar, with almost 100 additional trace nutrients, including many ions (including Ca^{2+}), some immunoglobulins (antibodies) in the form of IgA (Chapter 10) and growth factors, such as insulin-like growth factor-1 (IGF-1) and epidermal growth factor (EGF) (Chapter 46). **Colostrum**, the first secretion of the mammary glands after birth, is particularly rich in protein, but has a lower sugar concentration than mature milk. It also contains high levels of **antibodies** (Chapter 10) that provide the infant with basic immunological protection in the first days of life. At least four secretory processes are synchronized in the epithelial cells, **exocytosis**, **lipid synthesis** and secretion, **transmembrane secretion** of ions and water, and **transcytosis** of extra-alveolar proteins such as hormones, albumin and immunoglobulins from the interstitial spaces.

Hormonal control

Plasma prolactin levels rise steadily during pregnancy, but the lactogenic effects of the hormone are inhibited by the presence of progesterone and oestrogen, so that its main role during gestation is to promote mammary growth. Note however that progesterone and oestrogen are also essential during late pregnancy to stimulate duct and alveoli growth respectively, and without this pre-exposure the mammary glands will not respond to prolactin after birth. The loss of these placental steroids at term (Chapter 52) allows prolactin to exert its full effects on milk production, provided that cortisol and insulin are also present. **Placental lactogens**, which are similar to prolactin and are thought to bind to the same receptor, may contribute to mammary gland development during pregnancy, though their function in humans is not fully understood. Prolactin acts through a receptor linked to a Janus kinase–signal transduction and activation of transcription (JAK–STAT) system (Chapter 47) that activates the genes producing milk proteins and the synthetic enzymes for lactose and triglycerides. The production of nutrients is termed **galactopoiesis**. Prolactin also increases blood flow to the gland, and stimulates the delivery of nutrients into milk by exocytosis (proteins) or specific membrane transport systems (sugars, fats, antibodies); these actions are referred to as **lactogenesis**.

Prolactin is an unusual anterior pituitary hormone in that it is released **constitutively** (i.e. without a stimulus) from pituitary lactotrophs, and the primary control from the hypothalamus is inhibitory via **dopamine**, although other hypophysiotropic hormones may also be involved (Chapter 44; Fig. 53). After birth, the main stimulus that maintains prolactin release is **suckling**. Milk production thus continues for as long as the infant continues to feed from the mother. Prolactin inhibits luteinizing hormone (LH) release from the pituitary and maintains the mother in a low state of fertility until the infant is weaned. This is a useful mechanism for spacing births, but is not 100% effective in humans. Prolactin-secreting tumours of the pituitary render the patient infertile, but this can be overcome by the administration of the dopamine agonist bromocriptine, which inhibits prolactin release long enough for ovulation to occur. Prolactin is released in several conditions other than around birth: sleep, stress, eating and exercise are all associated with elevated plasma prolactin, although the exact function of this release is not yet known.

Milk let down reflex

Prolactin stimulates milk production, but another hormone is required to eject milk from the acini onto the surface of the nipple. Stimulation of areolar mechanoreceptors by suckling infants activates a neural pathway that ascends to the **paraventricular** and **supraoptic nuclei** of the hypothalamus via the lateral cervical nucleus of the brain stem. This pathway excites magnocellular neurones (Chapter 44) to secrete pulses of oxytocin into the blood at 2–10-min intervals. It is not certain how the suckling stimulus, which is continuous, is translated into episodic activity in oxytocin-releasing cells. The oxytocin pulses seem to arise from the simultaneous activation of all oxytocin neurones in both nuclei. The hormone is a potent stimulant of myoepithelial cells, which pump milk from the lactiferous sinuses out through the nipple and into the mouth of the infant. Milk let down encourages further suckling by the recipient, which leads to more oxytocin release, and so makes up another positive feedback system that operates until the infant is sated. This **milk ejection reflex** (Fig. 53) is also stimulated in response to the crying of infants as a result of psychological **conditioning**. However, the reflex is strongly inhibited by maternal stress, which is one of the most common causes of failure of lactation in new mothers. In animals, the release of oxytocin in the brain has been shown to facilitate maternal behaviour, but works only after pre-exposure to progesterone and oestrogens.

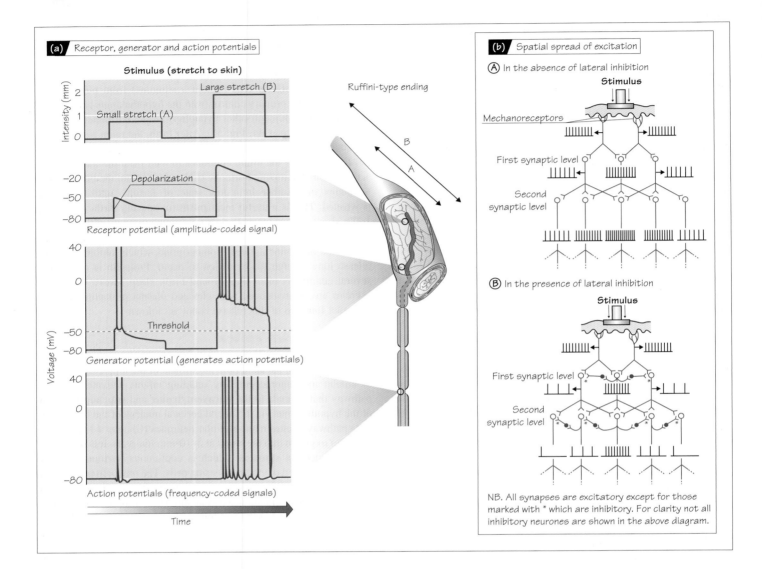

(a) Receptor, generator and action potentials

Stimulus (stretch to skin)

Intensity (mm)

Large stretch (B)

Small stretch (A)

Depolarization

Receptor potential (amplitude-coded signal)

Threshold

Generator potential (generates action potentials)

Voltage (mV)

Action potentials (frequency-coded signals)

Time

Ruffini-type ending

(b) Spatial spread of excitation

(A) In the absence of lateral inhibition

Stimulus

Mechanoreceptors

First synaptic level

Second synaptic level

(B) In the presence of lateral inhibition

Stimulus

First synaptic level

Second synaptic level

NB. All synapses are excitatory except for those marked with * which are inhibitory. For clarity not all inhibitory neurones are shown in the above diagram.

Physiology at a Glance, Third Edition. Jeremy P.T. Ward and Roger W.A. Linden.
 © 2013 John Wiley & Sons, Ltd. Published 2013 by John Wiley & Sons, Ltd.

Sensation and perception

The brain obtains its information about the external and internal environment and about the body's relation to the external environment by sensory experience emanating from sensory receptors (sense organs). There are a number of **common steps in sensory reception**: (i) a **physical stimulus** (i.e. touch, pressure, heat, cold, light, etc.); (ii) a **transduction process** (i.e. the translation of the stimulus into a code of action potentials); and (iii) a **response** (i.e. taking a mental note or triggering a motor reaction).

The **specialized nerve ending (sensory receptor)**, **afferent axon** and its **cell body**, together with the central synaptic connections in the spinal cord or brain stem, are known as **primary afferents**.

Information is transmitted to the brain in the form of action potentials. These action potentials carry this information in the form of **frequency-coded signals** and can signal the following information:

1 The **modality** (**specificity**) of the system. Such modalities include the 'five special senses': sight, hearing, balance, taste and smell. However, it is easy to list others. The skin itself not only senses pressure and touch, but also cold and warmth, vibration and pain (**somatosensation**). In addition, the body senses both the **external environment** and the **internal environment** (its own state). Examples are the sense of equilibrium (**balance**) and a knowledge of the relative positions of the limbs (**proprioception**). Other modalities that are related to information about the state of the body, and that are not directly apparent, are the senses that assess P_{CO_2} and P_{O_2}, blood pressure, and lung and stomach stretch receptors, the so-called **interoceptors**. Each modality can often be subdivided into further divisions of **quality**, i.e. in the case of taste (sweet, sour, salt, bitter and umami), light (red, green and blue) and hearing (tonal pitches).

2 The **intensity** (**quantity**) of the stimulus (Fig. 54a). The quantity of a sensory impression corresponds to the strength of the stimulus. As the stimulus strength increases, so does the amplitude of the receptor potential (**amplitude-coded signal**) and, when this eventually reaches a **threshold**, it causes action potentials that increase in their frequency of firing as the receptor potential rises (**temporal** or **frequency coding**). Another way in which the strength of the signal is coded is by increasing the number of afferent fibres that are activated (**spatial** or **recruitment coding**).

3 The **duration** of the stimulus. Many receptors will continue to fire impulses as long as the stimulus is applied; others will signal when a stimulus is applied and when a stimulus is removed. However, in most cases, even if a stimulus persists (e.g. constant touch to the skin), the sensation/perception of it wanes. This involves a process called **adaptation**. **Adaptation** occurs at all stages of the transformation of the stimulus: in the transduction process, in the conductance mechanism of the receptor potential, in the synaptic transmission from a secondary sensory cell and in the generation of the action potential. It can also be a function of the central nervous system (CNS) itself once the action potentials reach that far.

4 The **localization** and **resolution** (**acuity**) of the stimulus. The sensory system detects the location of a stimulus, and its fine detail. Both depend on the spacing of receptors (better localization and acuity occur with greater receptor density). The **receptive field** of a sensory neurone itself (sometimes called the **receptor field**) is the area of sensory surface from which that neurone receives an input. Receptor neurones converge onto second-order neurones (usually in the CNS), and then to third- and higher-order neurones. These transitions are made in relay nuclei. The receptor field of the primary receptor is usually a small excitatory area. The receptive field of the second- or higher-order neurone is larger and more complex (because of both convergence and divergence, and excitatory and inhibitory pathways).

The net result is **sensation** and, when interpreted at a conscious level in the light of experience, this becomes **perception**.

Sensory pathways

The coded signals from each of the sensory receptors are relayed to the CNS by peripheral and cranial nerves. Each modality is associated with specific nerves and pathways, e.g. gustatory information is transmitted via facial and glossopharyngeal nerves, and the somatosensory system is transmitted via the dorsal column–medial lemniscal system for the larger afferent fibres (Aα and Aβ) and the anterolateral system (anterior and lateral spinothalamic tracts) for the smaller afferent fibres (Aδ and C). Each sensory system has its unique pathway into and through the CNS to eventually provide an input into the thalamus. The thalamus, in turn, provides an input to the cortex. Each sensory system projects to a specific area of the primary sensory cortex which is primarily concerned with the analysis of the sensory information, and these neurones, in turn, project to the secondary sensory cortex in which more complex processing occurs. There are further projections to associated areas, such as the posterior parietal, prefrontal and temporal cortices, which can again project to the limbic and motor systems. The latter systems are involved in the processing of the sensory information, leading to responses such as complex behavioural and motor responses.

Lateral inhibition. Figure 54b shows a neural network comprising two mechanoreceptors in the skin and their associated neurones at the next two synaptic levels. The two receptors are each excited equally by a stimulus applied between them. The **divergent** and **convergent connections** seem to impose an avalanche-like spread of excitation at progressively higher levels of the CNS. Pinpoint stimulation appears to lead to an enlarged, less precise and more diffuse representation at each successive synaptic level (Ⓐ). However, this situation is encountered only under pathological conditions (e.g. strychnine poisoning, which blocks inhibiting synapses in the CNS). Inhibition normally prevents the spread of excitation by a phenomenon called **lateral inhibition** (Ⓑ). At each synaptic relay, each excitatory neurone exerts an inhibitory effect by exciting **inhibitory interneurones**. The neurone with the greatest input (the one in the middle) imposes the strongest inhibition on those on either side of it. Lateral inhibition has been shown to exist at all levels of sensory systems: in the dorsal horn of the spinal cord, in the dorsal column nuclei, in the thalamus and in the cortex, as well as in the visual system. The result is an increased spatial sharpening in the CNS of the representation of the distant peripheral stimulus on moving through the synaptic levels.

Descending inhibition. In practically all sensory systems, higher centres can also exert inhibitory effects on all those at lower levels. Such **central inhibition** can act at a point as far peripheral as the receptor or at the afferent ending in the spinal cord. Like lateral inhibition, descending inhibition can be considered to function as a means of **regulating the sensitivity of the afferent transmission channels**.

The types of synaptic mechanism described above indicate that there is great flexibility in the sensory pathways, and that they are not as hard-wired as many pathway diagrams suggest.

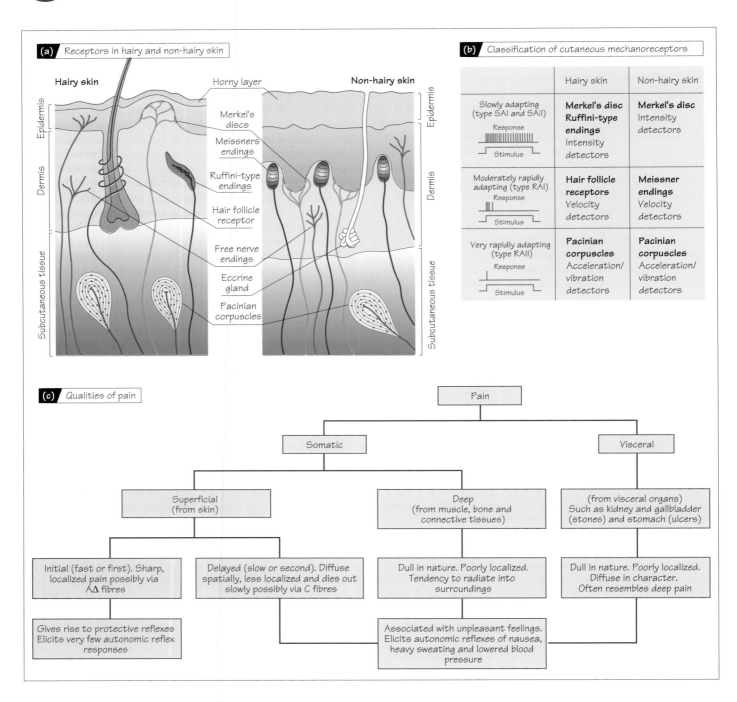

(a) Receptors in hairy and non-hairy skin

Hairy skin

Non-hairy skin

Horny layer

Merkel's discs

Meissners endings

Ruffini-type endings

Hair follicle receptor

Free nerve endings

Eccrine gland

Pacinian corpuscles

Epidermis / Dermis / Subcutaneous tissue

(b) Classification of cutaneous mechanoreceptors

	Hairy skin	Non-hairy skin
Slowly adapting (type SAI and SAII) Response / Stimulus	**Merkel's disc Ruffini-type endings** Intensity detectors	**Merkel's disc** Intensity detectors
Moderately rapidly adapting (type RAI) Response / Stimulus	**Hair follicle receptors** Velocity detectors	**Meissner endings** Velocity detectors
Very rapidly adapting (type RAII) Response / Stimulus	**Pacinian corpuscles** Acceleration/ vibration detectors	**Pacinian corpuscles** Acceleration/ vibration detectors

(c) Qualities of pain

Pain
- Somatic
  - Superficial (from skin)
    - Initial (fast or first). Sharp, localized pain possibly via AΔ fibres
      - Gives rise to protective reflexes Elicits very few autonomic reflex responses
    - Delayed (slow or second). Diffuse spatially, less localized and dies out slowly possibly via C fibres
  - Deep (from muscle, bone and connective tissues)
    - Dull in nature. Poorly localized. Tendency to radiate into surroundings
    - Associated with unpleasant feelings. Elicits autonomic reflexes of nausea, heavy sweating and lowered blood pressure
- Visceral
  - (from visceral organs) Such as kidney and gallbladder (stones) and stomach (ulcers)
    - Dull in nature. Poorly localized. Diffuse in character. Often resembles deep pain

Physiology at a Glance, Third Edition. Jeremy P.T. Ward and Roger W.A. Linden.
 © 2013 John Wiley & Sons, Ltd. Published 2013 by John Wiley & Sons, Ltd.

The **sensory receptor** is a specialized cell. In mammals, receptors fall into five groups: **mechanoreceptors**, **thermoreceptors**, **nociceptors**, **chemoreceptors** (Chapter 56) and **photoreceptors** (Chapter 57). There is further specialization within these groups. Each receptor responds to one stimulus type; this property is called the **specificity of the receptor**. The stimulus that is effective in eliciting a response is called the **adequate stimulus**.

Transduction processes. Some receptors consist of a nerve fibre alone (e.g. free nerve endings), others consist of a specialized accessory structure (e.g. olfactory receptors, Pacinian corpuscles), and others are more complex and consist of a specialized receptor cell which synapses with a neurone, in other words a secondary sensory cell (e.g. gustatory receptors and Merkel's discs).

Mechanoreceptors. These are found all over the body. Those in the skin have three main qualities: pressure, touch and vibration (or acceleration) (Fig. 55a,b). When the responses to constant stimuli are studied in the various receptors, the receptors can be divided into three types on the basis of their adaptive properties: **slowly adapting receptors** that continue to fire action potentials even when the pressure is maintained for a long period (e.g. **Ruffini's endings, tactile discs, Merkel's discs**); **moderately rapidly adapting receptors** that fire for about 50–500 ms after the onset of the stimulus, even when the pressure is maintained (e.g. **hair follicle receptors, Meissner's corpuscles**); and **very rapidly adapting receptors** that fire only one or two impulses (e.g. **Pacinian corpuscles**) (Fig. 55b). These three types of receptor are examples of receptors in the skin that detect **intensity**, **velocity** and **vibration** (or **acceleration**), respectively.

Free nerve endings. Each skin nerve, in addition to the large myelinated afferents, contains a large number (over 50% of the fibres) of **smaller myelinated and unmyelinated (Aδ and C) axons**. Some of the C fibres are, of course, efferent postganglionic sympathetic fibres. However, a large number of the remaining fibres are afferents that terminate in **free nerve endings** and not in corpuscular structures (Fig. 55a). Many of these are **thermoreceptors** or **nociceptors**.

Thermoreceptors. Thermoreceptors mediate the sensations of **cold** and **warmth**. In the skin of humans, there are **specific cold and warm points** at which only the sensation of cold or warmth can be elicited. These are specific cold and warmth receptors; however, they share the following characteristics: (i) **maintain discharge** at constant skin temperature, with the discharge rate proportional to the skin temperature (static response); (ii) have **small receptive fields** ($1 \, mm^2$ or less), each afferent fibre supplying only one or a few warm or cold points; and (iii) serve not only as **sensors for the conscious sensation of temperature**, but also participate (together with temperature sensors in the hypothalamus and spinal cord) in the **thermoregulation** of the body.

Nociceptors and pain. Pain differs from the other sensory modalities with regard to the kind of information it conveys. It informs us of a threat to our bodies when it is activated by noxious (tissue-damaging) stimuli. **Nociception** is defined as the **reception, conduction and central processing of noxious signals**. This term is used to make a clear distinction between these 'objective' neuronal processes and the 'subjective' sensation of **pain**, which is defined as an **unpleasant sensory and emotional experience associated with actual or potential damage, or described in terms of such damage**.

Nociceptors are found in the skin, visceral organs and muscle (cardiac and skeletal), and are associated with blood vessels. The **qualities of pain** are divided into **somatic** and **visceral**. If somatic pain is derived from the skin, it is called **superficial pain**, and, if from muscle, bone joints or connective tissue, it is called **deep pain**. If superficial pain is produced by piercing the skin with a needle, the subject feels a sharp pain; this easily localized sensation fades away rapidly when the needle is removed. This sharp, localized **initial pain** (also called **first** or **fast** pain) is often followed, particularly at high stimulus intensities, by **delayed pain** (also called **second** or **slow** pain), which has a dull (or burning) character with a delay of about 1 s. This delayed pain is more diffuse spatially, dies out slowly and is not so easily localized. **Deep pain** is dull in nature, poorly localized and has a tendency to radiate into the surroundings.

The responses of the body in terms of both the distress and suffering and the autonomic and motor responses to pain depend on the quality of pain (Fig. 55c). **Delayed pain** and **deep pain** are accompanied by a **feeling of unpleasantness**, and often elicit autonomic reflexes of **nausea, heavy sweating** and **lowered blood pressure**. **Initial pain** gives rise, by contrast, to **protective reflexes**, i.e. flexor withdrawal reflex. **Visceral pain** (pain from organs such as the kidney, stomach and gallbladder) tends to be dull and diffuse in character and resembles deep pain.

Histologically, the nociceptors are **free nerve endings** attached to either **Aδ fibres** or **C fibres**. It has been proposed that, in the case of superficial pain, the transmission of **initial (fast) pain** is via **Aδ** fibres, whereas **delayed (slow) pain** is signalled by the smaller **C** fibres. The time difference between initial (fast) pain and delayed (slow) pain appears to be explained by the difference in the conduction velocities of the fibres concerned.

Inhibitory influences. Like all other sensory inputs, the nociceptive afferent influx is exposed to inhibitory influences at the receptor, on its way to and through the spinal cord and in the higher levels of the central nervous system. Many of the modern treatments elicit or enhance these inhibitory processes, pharmacologically using drugs, physically using cold or warm wrappings, short-wave radiation, massage and exercise, and by the electrical stimulation of certain structures, including peripheral nerves. **Acupuncture** and **transcutaneous electrical nerve stimulation** (TENS) may possibly depend on the activation and maintenance of inhibitory processes. Naturally occurring **endorphins**, **enkephalins** and **dynorphins** are thought to contribute to these processes. These are endogenous, pain-controlling opiates produced by the body that attach to the specific opiate receptors, so as to inhibit the sensation of pain without affecting the other sensory modalities.

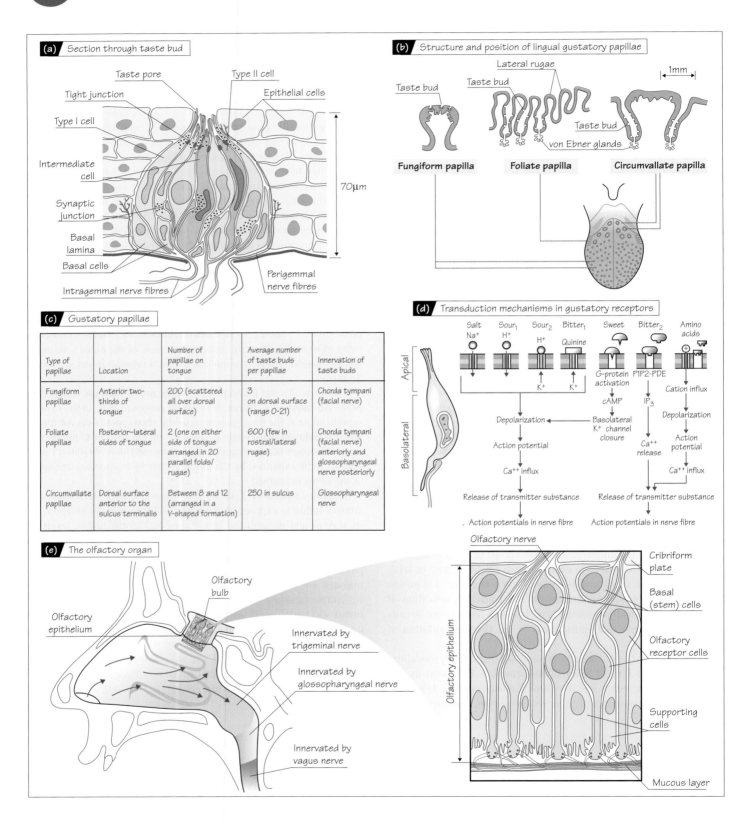

(a) Section through taste bud

(b) Structure and position of lingual gustatory papillae

Fungiform papilla — Foliate papilla — Circumvallate papilla

(c) Gustatory papillae

Type of papillae	Location	Number of papillae on tongue	Average number of taste buds per papillae	Innervation of taste buds
Fungiform papillae	Anterior two-thirds of tongue	200 (scattered all over dorsal surface)	3 on dorsal surface (range 0-21)	Chorda tympani (facial nerve)
Foliate papillae	Posterior–lateral sides of tongue	2 (one on either side of tongue arranged in 20 parallel folds/rugae)	600 (few in rostral/lateral rugae)	Chorda tympani (facial nerve) anteriorly and glossopharyngeal nerve posteriorly
Circumvallate papillae	Dorsal surface anterior to the sulcus terminalis	Between 8 and 12 (arranged in a V-shaped formation)	250 in sulcus	Glossopharyngeal nerve

(d) Transduction mechanisms in gustatory receptors

(e) The olfactory organ

Physiology at a Glance, Third Edition. Jeremy P.T. Ward and Roger W.A. Linden.

© 2013 John Wiley & Sons, Ltd. Published 2013 by John Wiley & Sons, Ltd.

The so-called special senses comprise the sensations of **taste, smell, vision, hearing** and **balance**. The receptors involved in taste and smell are **chemoreceptors**, those in vision are **photoreceptors**, and those in hearing and balance are **mechanoreceptors**.

The sensations of **taste** and **smell** are two modalities of sense that are very closely related. What the layperson calls 'taste' is really a combination of taste and smell, and probably a number of other modalities. Taken together, a better term would be **flavour**. The modalities of flavour are **taste** (**gustation**), **smell** (**olfaction**), **touch** (**texture**), **temperature** (**thermoreception**) and **common chemical sense** (**chemoreception**).

Gustation

Taste buds (the **gustatory end organs**) (Fig. 56a) are found in the tongue, soft palate, pharynx, larynx and epiglottis, and are unevenly distributed around these regions. Those in the tongue are associated with three of the four types of papillae (**fungiform, foliate** and **circumvallate**) (Fig. 56b). Those associated with the other oral tissues are found on the smooth epithelial surfaces. The different papillae occupy specific areas of the tongue. Their associated taste buds are innervated by either the **glossopharyngeal** (**IX**) **nerve** (posterior one-third of the tongue) or the **chorda tympani branch** of the **facial** (**VII**) **nerve** (anterior two-thirds of the tongue). In humans, the number of taste buds varies considerably: on average in the range 2000–5000, but can be as low as 500 or as high as 20000 (Fig. 56c). Each taste bud is made up of 50–150 **neuroepithelial cells** arranged in a compact, pear-shaped structure (**intragemmal cells**). There is general agreement that there are four types of intragemmal cells: **basal, type I (dark cells), intermediate** and **type II (light cells)**. Each taste bud comprises a dynamic system in which there is a rapid turnover of cells within each bud. The **lifespan** of an individual receptor cell is about **10 days**. There is a small opening in the surface, the **taste pore**, where the cells have access to the gustatory stimuli (Fig. 56a).

Since Aristotle (384–322 BC), people have tried to categorize taste into **primary** or **basic qualities of taste**. The four qualities that have stood the test of time are **sweet, sour, salt** and **bitter**, with a fifth categorized by the taste of monosodium glutamate (**umami**). The mechanisms involved in the process of transduction of the signals that eventually produce these basic sensations of taste are complex. In common with many other receptor cells, gustatory receptor cells use **specifically localized ion channel and receptor sites** for transduction. Unlike many other receptor cells, there is **no single membrane transduction event** and the different basic tastes utilize **different ionic mechanisms** (Fig. 56d). Most gustatory stimuli are water soluble and non-volatile and are either already dissolved or are dissolved in saliva during mastication.

The **common chemical sense** has been defined as the sensation caused by the stimulation of **epithelial** or **mucosal free nerve endings** by chemicals. Evidence suggests that these are **polymodal nociceptors** and that, in the mouth, the major contributor to this sense is the **trigeminal** (**V**) **nerve**. The trigeminal innervates almost all regions of the mouth, including the floor of the mouth, the tongue, the hard and soft palate, and the mucosa of the lips and cheek. These nerve endings are stimulated by a number of different chemicals, such as menthol, peppermint, and capsaicin and piperine (found in chilli peppers and black peppers, respectively).

Olfaction

The human olfactory organ, the **olfactory epithelium** or **mucosa**, is a sheet of cells, 100–200-μm thick, situated high in the back of the nasal cavity and on the thin bony partition (the **central septum**) of the nasal passage. The olfactory system responds to airborne, volatile molecules that gain access to the olfactory epithelium with the in-and-out air flow through and behind the nose. The odour molecules are distributed over the receptor sheet in an irregular pattern by the turbulence of the air flow set up by the turbinate bones (Fig. 56e).

The olfactory epithelium contains specialized, elongated nerve cells (**olfactory receptors**) (Fig. 56e). These cells have very thin fibres that run upwards in bundles through perforations in the skull (the **cribriform plate**) above the roof of the nasal cavity. These bundles of nerves constitute the **olfactory** (**I**) **nerve**. They extend only a very short distance, ending in the **olfactory bulbs**, a pair of swellings underneath the frontal lobes. The other end of each olfactory receptor, pointing down into the nasal cavity, is extended into a long process, ending in a knob carrying several hairs (**cilia**) between 20 and 200 μm in length. These cilia are bathed in a thin (35-μm thick) layer of **mucus**, secreted by specialized cells in the olfactory epithelium, in which the molecules of odorous substances dissolve. The molecules diffuse through the surface layer of mucus and stimulate the olfactory receptors. **Hydrophilic** (water-soluble) molecules dissolve readily in the mucus, but the diffusion of less soluble molecules is assisted by 'odour-binding proteins' in the mucus, which are also thought to assist in removing odour molecules from the receptor cells. The mucus layer moves across the surface of the **olfactory mucosa** at 10–60 mm/min towards the **nasopharynx**. This flow of mucus also assists in the removal of odours after they have been sensed. In the membrane of the cilia are **olfactory receptor proteins**, which interact with the smelly molecules, and initiate a cascade reaction inside the cell that leads to a change in the rate of impulses.

Humans are able to **distinguish 10000 or more different odours**. Individual olfactory receptor neurones fire off spontaneously at between 3 and 60 impulses per second. When stimulated with particular odours, they increase their firing frequency. Each **receptor cell** responds, although not equally, to many different types of odour.

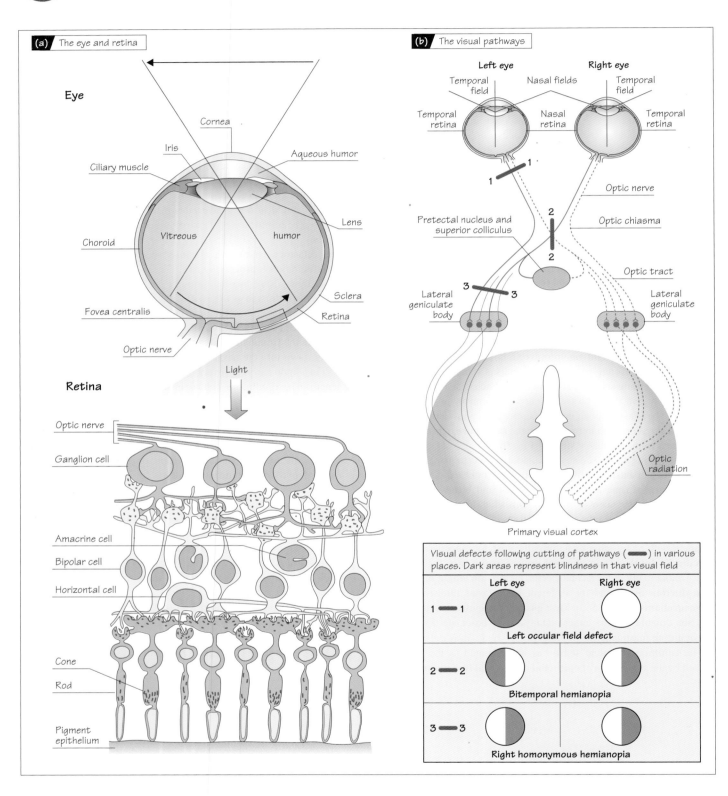

(a) The eye and retina

Eye

Cornea
Iris
Ciliary muscle
Aqueous humor
Lens
Choroid
Vitreous   humor
Fovea centralis
Sclera
Retina
Optic nerve

Light

Retina

Optic nerve
Ganglion cell
Amacrine cell
Bipolar cell
Horizontal cell
Cone
Rod
Pigment epithelium

(b) The visual pathways

Left eye   Right eye
Temporal field   Nasal fields   Temporal field
Temporal retina   Nasal retina   Temporal retina
1
1
Optic nerve
2
Pretectal nucleus and superior colliculus
2
Optic chiasma
Optic tract
3   3
Lateral geniculate body   Lateral geniculate body
Optic radiation

Primary visual cortex

Visual defects following cutting of pathways (▬▬) in various places. Dark areas represent blindness in that visual field

	Left eye	Right eye
1 ▬▬ 1	●	○
	Left occular field defect	
2 ▬▬ 2	◑	◐
	Bitemporal hemianopia	
3 ▬▬ 3	◐	◐
	Right homonymous hemianopia	

Physiology at a Glance, Third Edition. Jeremy P.T. Ward and Roger W.A. Linden.
 © 2013 John Wiley & Sons, Ltd. Published 2013 by John Wiley & Sons, Ltd.

Vision in humans involves the detection of a very narrow band of light ranging from about **400 to 750 nm** in wavelength. The shortest wavelengths are perceived as blue and the longest as red. The eye contains **photoreceptors** that detect light, but, before the light hits the receptors responsible for this detection, it has to be focused onto the retina (200 μm thick) by the cornea and the lens (Fig. 57a).

The photoreceptors can be divided into two distinct types called **rods** and **cones**. **Rods** respond to **dim light** and **cones** respond in **brighter conditions and can distinguish red, green or blue light**. The rods and cones are found in the deepest part of the retina, and light has to travel through a number of cellular layers to reach these photoreceptors. Each photoreceptor contains molecules of the **visual pigments** (rods: **rhodopsin**; cones: **erythrolabe (red)**, **chlorolabe (green)** and **cyanolabe (blue)**); these absorb light and trigger receptor potentials which, unlike other receptor systems, lead to a **hyperpolarization** of the cell and not depolarization.

The layers between the retina surface and the receptor cells contain a number of excitable cells, called **bipolar, horizontal, amacrine** and **ganglion cells**. The **ganglion cells** are the neurones that transmit impulses to the rest of the central nervous system (CNS) via axons in the **optic nerve**. These cells are excited by the vertical **bipolar interneurones** which lie between the receptor cells and the ganglion cells. In addition, this complex structure also contains two groups of interneurones (**horizontal** and **amacrine cells**) that function by exerting their influence in a horizontal manner, by causing lateral inhibition on surrounding synaptic connections between receptor cells and bipolar cells, and bipolar cells and ganglion cells, respectively (Fig. 57a).

Each eye contains approximately **126 million photoreceptors (120 million rods and 6 million cones)** but only **1.5 million ganglion cells**. This means that there is a substantial amount of convergence of receptor and bipolar cells onto ganglion cells, but this is not uniform across the retina. At the periphery, there is a large amount of convergence, but, in the region of greatest visual clarity (the **fovea centralis**), there is a 1:1:1 connectivity between a single cone receptor cell, a single bipolar cell and a single ganglion cell. The fovea region has a very high density of cones and very few rods, whereas there is a more even distribution of rods and cones in the other regions of the retina.

Each ganglion cell responds to changes in light intensity over a limited area of the retina, rather than to a stationary light stimulus. This limited area is called the **receptive field** of the cell and corresponds to the group of photoreceptors that has synaptic connections with that particular ganglion cell. Ganglion cells are usually spontaneously active. Approximately half of the ganglion cells in the retina respond with a decrease in firing of their impulses when the periphery of their receptive field is stimulated by light, and increase their firing rate when the centre of the receptive field is lit up (the **ON-centre cells**); the other half increase their firing rate when the periphery is illuminated and decrease their firing rate when the central receptors are stimulated (the **OFF-centre cells**). This allows the output of the retina to signal the relative brightness and darkness of each area being stimulated within the visual field.

The ganglion cells are further subdivided into two main groups: P cells and M cells. P cells receive the central parts of their receptive fields from one or possibly two (but never all three) types of colour-specific cone, whereas M cells receive inputs from all types of cone. M cells are therefore not colour selective, but sensitive to contrast and movement of images on the retina. The division of P and M cells appears to be maintained throughout the visual pathway and they are involved in visual perception.

The optic nerves from the two eyes join at the base of the skull at a structure called the **optic chiasma** (Fig. 57b). Approximately half of each of the optic nerve fibres cross over to the contralateral side; the other half remain on the ipsilateral side and are joined by axons crossing from the other side. The axons of the ganglion cells from the **temporal region** of the retina of the left eye and the **nasal region** of the retina of the right eye proceed into the left optic tract, whereas the axons from the ganglion cells in the **nasal part** of the left eye and the **temporal part** of the right eye form the right optic tract. The neurones that make up the **optic tract** connect to the first relay stations in the pathway: the **lateral geniculate bodies**, the **superior colliculus** and the **pretectal nucleus** of the **brain stem**. Those fibres that synapse in the superior colliculus and the pretectal nucleus are involved in visual reflexes and orientating responses. A small number of fibres also branch off at this point to synapse in the **suprachiasmatic nucleus**, which is concerned with the body clock and circadian rhythms within the body. However, the bulk of the neurones reach the **lateral geniculate nucleus** in the **thalamus**. Each nucleus contains six cellular layers and the information from the two eyes remains separate, each group of fibres synapsing in three of the layers. The **M ganglion cells** terminate in the lower two layers (called **magnocellular** because the cells are relatively large in these layers). Cells in the magnocellular layers are sensitive to contrast and motion, but not colour. The **P ganglion cells** synapse in the upper four layers of the lateral geniculate nucleus (two for each eye), called the **parvocellular** layers. These layers contain relatively small cells which transmit information about colour and fine detail. The fibres from the lateral geniculate nucleus fan backwards and upwards in a bundle (called the **optic radiation**) through the **parietal** and **temporal lobes** to an area of the cerebral cortex called the **primary visual cortex**. Each cortical cell receives inputs from a limited number of cells in the lateral geniculate nucleus, and therefore has its own receptive field or patch of retina to which it responds.

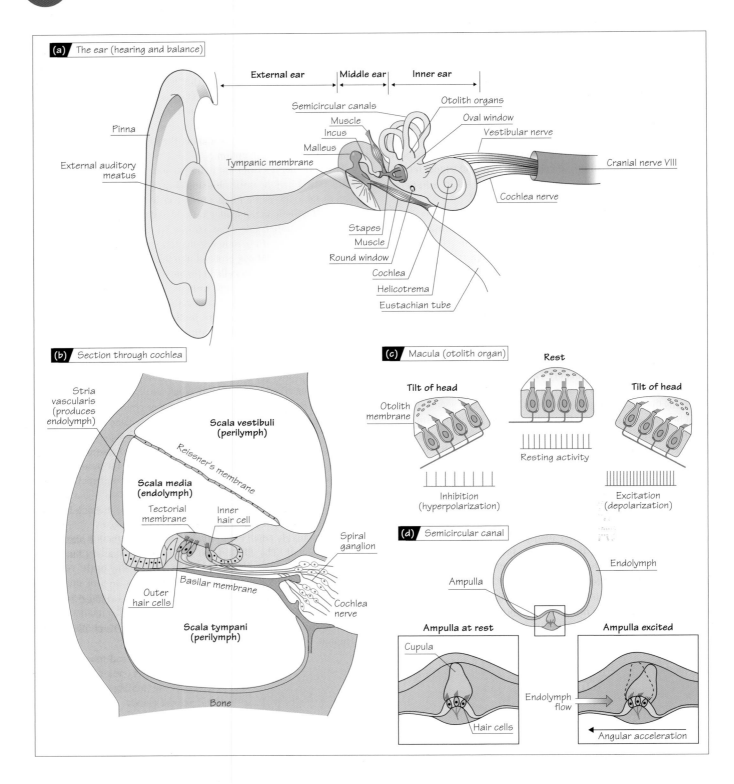

Physiology at a Glance, Third Edition. Jeremy P.T. Ward and Roger W.A. Linden.

126 © 2013 John Wiley & Sons, Ltd. Published 2013 by John Wiley & Sons, Ltd.

Hearing

The young healthy human can detect sound wave frequencies of between **40 Hz** and **20 kHz**, but the upper frequency limit declines with age. When sound waves reach the ear, they pass down the **external auditory meatus (the external ear)** to the **tympanic membrane** that vibrates at a frequency and strength determined by the magnitude and pitch of the sound. The vibration of the membrane causes three ear **ossicles (malleus, incus** and **stapes)** in the **middle ear** (an air-filled cavity) to move, which, in turn, displaces fluid within the **cochlea (the inner ear)** as the foot of the stapes moves the **oval window** at the base of the cochlea. This mechanical link prevents the incoming sound energy from being reflected back, and the ossicles improve the efficiency with which the sound energy is transferred from the air to the fluid. Small muscles are attached to the ossicles and contract reflexly in response to loud sounds, thereby dampening the vibration and attenuating the transmission of the sound (Fig. 58a).

The inner ear includes the **cochlea** and also the **vestibular organs** responsible for balance (see later). The receptors involved in both hearing and balance are specialized mechanoreceptors called **hair cells**. Projecting from the apical surface of the hair cell is a bundle of over 100 small hair-like structures called **stereocilia** and a larger stereocilium called the **kinocilium**. Deflection of the stereocilia towards the kinocilium leads to a potential change in the cell (depolarization), the release of a transmitter substance from the base of the hair cell, and activation of the nerve fibres that convey impulses to the higher centres of the brain.

The cochlea comprises a coiled tube of about 3 cm in length (Fig. 58b), with three tubular canals running parallel to one another (**scala vestibuli, scala media** and **scala tympani**).

The scala vestibuli and the scala tympani contain **perilymph** (which is similar to extracellular fluid in composition), and the scala media contains **endolymph** (similar in composition to intracellular fluid). The scala vestibuli and scala tympani are joined at the tip of the coil (the **helicotrema**); at the base of the scala vestibuli is the **oval window** and at the base of the scala tympani is the **round window**, separating the fluid of the inner ear from the air in the middle ear.

The scala media lies between the two perilymph-filled canals; the boundary between it and the scala vestibuli is called **Reissner's membrane**, and the boundary between it and the scala tympani is called the **basilar membrane**. On top of the basilar membrane sits the **organ of Corti** in which the hair cells are situated. There are around 15 000 hair cells distributed in rows along the basilar membrane. There are two types of hair cell: the **inner hair cells** which form a single row and the more numerous **outer hair cells** arranged in three rows. The hair cells are ideally placed to detect small amounts of movement of the basilar membrane. Because of the changing width of the basilar membrane, high-frequency sounds maximally displace the membrane at the base of the cochlea and low-frequency sounds maximally displace the membrane at the apical end of the cochlea.

The auditory signals are relayed through a complex series of nuclei in the brain stem and the thalamus, eventually reaching the **primary auditory cortex** in the temporal lobe of the cerebral cortex.

Balance

The system associated with balance is called the **vestibular system** and is not only involved with balance, but also postural reflexes and eye movements.

As mentioned earlier, the receptors involved in the vestibular system are hair cells. These hair cells are found in the inner ear in close proximity to the cochlea in two **otolith organs** called the **utricle** and **saccule**, and in a structure called the **ampulla** found in the three semicircular canals. The otolith organs are primarily involved in the detection of **linear motion** and **static head position**, and the semicircular canals in the detection of **rotational movements of the head**.

The four otolith organs (two on each side) each contain a structure called the **macula** which comprises a number of hair cells (Fig. 58c). With the head erect, the macula in each utricle is orientated horizontally and that in each saccule is orientated vertically. The base of each macula contains hair cells whose stereocilia project into a gelatinous mass called the **otolith membrane**. When the head is tilted, the force of gravity displaces the otolith membrane, thereby bending the stereocilia. The nerve fibres innervating the hair cells are spontaneously active: displacement in one direction increases firing and displacement in the opposite direction decreases firing of the neurones. The **utricle** sends signals representing **forwards and backwards movements** and the **saccule** conveys information about **vertical movements**.

The semicircular canals each contain an organ called the **ampulla** (Fig. 58d). They respond to **rotational movement of the head**, and the plane of each canal is perpendicular to the other two, so that, between all six (three on each side), they provide information relating to the rotational acceleration of the head during movement around any axis. Each canal contains endolymph and the ampulla comprises hair cells in which the stereocilia project into a gelatinous mass, with the same specific gravity as the endolymph, called the **cupula**. During acceleration in the plane of a particular canal, the endolymph tends to remain stationary because of inertia. The movement displaces the stereocilia and stimulation of the associated nerve fibres occurs. Again, movement in one direction increases firing of the nerves and movement in the opposite direction causes a decrease in firing. Vestibular afferent fibres from the auditory (VIII) nerve have their cell bodies in the **vestibular ganglion** and terminate in one of **four vestibular nuclei in the medulla**. These nuclei also receive inputs from neck muscle receptors and the visual system. They then project to a number of areas of the central nervous system, including the **spinal cord, thalamus, cerebellum** and **oculomotor nuclei**, where they are involved in posture, gait and eye movements. They also project to the **primary somatosensory cortex** and to the **posterior parietal cortex**.

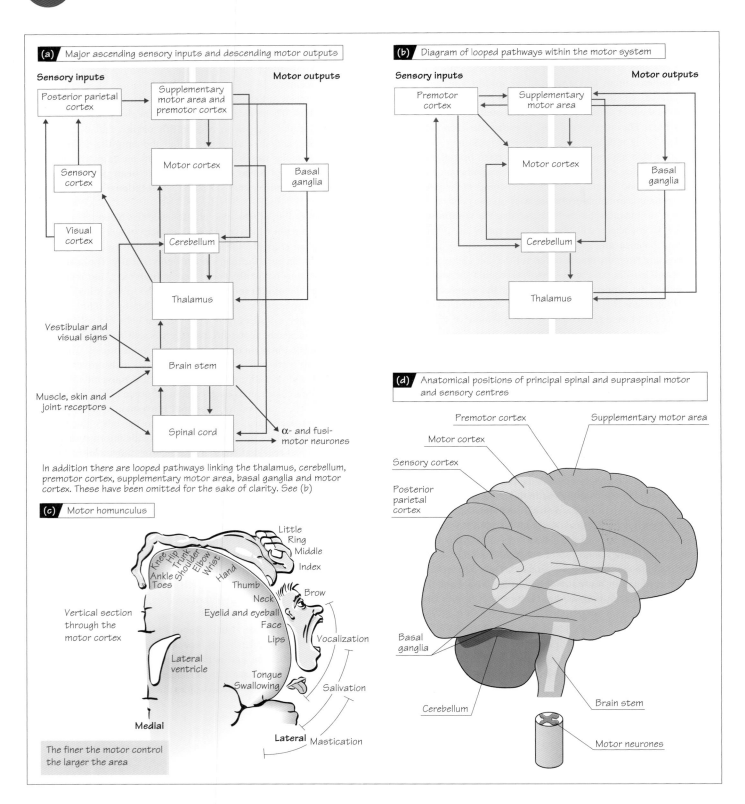

(a) Major ascending sensory inputs and descending motor outputs

Sensory inputs   Motor outputs

Posterior parietal cortex

Supplementary motor area and premotor cortex

Sensory cortex

Motor cortex

Basal ganglia

Visual cortex

Cerebellum

Thalamus

Vestibular and visual signs

Brain stem

Muscle, skin and joint receptors

Spinal cord

α- and fusi-motor neurones

In addition there are looped pathways linking the thalamus, cerebellum, premotor cortex, supplementary motor area, basal ganglia and motor cortex. These have been omitted for the sake of clarity. See (b)

(b) Diagram of looped pathways within the motor system

Sensory inputs   Motor outputs

Premotor cortex

Supplementary motor area

Motor cortex

Basal ganglia

Cerebellum

Thalamus

(c) Motor homunculus

Little
Ring
Middle
Index
Knee Hip Trunk Shoulder Elbow Wrist Hand
Ankle
Toes
Thumb
Neck
Brow
Vertical section through the motor cortex
Eyelid and eyeball
Face
Lips
Vocalization
Lateral ventricle
Tongue
Swallowing
Salivation
Medial
Lateral Mastication

The finer the motor control the larger the area

(d) Anatomical positions of principal spinal and supraspinal motor and sensory centres

Premotor cortex
Supplementary motor area
Motor cortex
Sensory cortex
Posterior parietal cortex
Basal ganglia
Cerebellum
Brain stem
Motor neurones

Physiology at a Glance, Third Edition. Jeremy P.T. Ward and Roger W.A. Linden.

 © 2013 John Wiley & Sons, Ltd. Published 2013 by John Wiley & Sons, Ltd.

Motor control

Motor control is defined as the control of movements by the body. These movements can be both influenced and guided by the many sensory inputs that are received, or can be triggered by sensory events. They can also be triggered by the need to move using internal mechanisms. The major division of the body into sensory and motor functions is artificial, because almost all motor areas in the central nervous system (CNS) receive sensory inputs.

The organization and physiology of motor systems have been represented as a number of **hierarchical structures**, but these must be viewed with caution, as they are again artificial and, by necessity, oversimplified.

Figure 59a shows the major ascending sensory inputs and descending motor outputs, and Figure 59b shows the main looped pathways within the CNS.

Voluntary movements can be summarized as follows. Exactly where the idea of a movement is initiated is unknown, but it is thought to be in the areas of the cortex other than the primary sensory or primary motor cortices (the **association cortex**) and possibly the **basal ganglia**. At this stage, sensory information relating to the intended movement is analysed in the **posterior parietal cortex**. This sensory information mainly comes from the **visual** and **sensory cortex**.

The posterior parietal cortex activates the **supplementary motor area** and the **premotor cortex**. This excitation also causes the **basal ganglia loop** and the **cerebrocerebellar loop** to be excited and to lead to a degree of amplitude setting and coordination of the activity. The supplementary motor area and the premotor cortex then initiate activity in the **motor cortex**. In addition, the premotor cortex initiates, via the **anterior corticospinal tract** and the connections to the brain stem **ventromedial pathways**, any postural adjustments needed for the movement.

The motor cortex, via the lateral **corticospinal** and **corticorubrospinal tracts**, then initiates the activity of the muscles. This activity is due to the excitation of both α- and **fusi-motor neurones**. During this movement, there is continuous feedback from **receptors** in the **joints**, **muscles** and **skin**, which can lead to fine adjustments via local **spinal** and **brain stem reflexes**. Furthermore, there is often **visual feedback** which can modulate the motor outputs at the cortical and cerebellar levels. Modulations of the activity at all levels continue throughout the voluntary movement.

Figure 59d shows the anatomical sites of the principal motor and sensory centres, and Figure 59c shows the relative size of the areas in the motor cortex represented by the different parts of the body (the motor homunculus).

The term **upper motor neurones** refers to those neurones that are wholly in the CNS motor pathways. These descending motor pathways are divided into the **pyramidal tracts**, which originate in the cerebral cortex, and the **extrapyramidal tracts**, which originate in the brain stem. The **pyramidal tracts** descend through the **internal capsule** and terminate in the brain stem. One small group of fibres (the **corticobulbar tract**) terminates on cranial motor nuclei and is involved in controlling eye, facial and masticatory muscles. Another larger group of fibres (the **corticospinal tract**) descends directly from the cortex to the grey matter of the spinal cord but, as it passes through the brain

stem, it divides into two. Approximately **85%** of the fibres cross over the midline (**decussate**) and descend as the **lateral corticospinal tract**, terminating directly onto the α- and fusi-motor neurones. Some of the fibres do not terminate directly onto the motor neurones but excite interneurones instead. These interneurones can be either excitatory or inhibitory in nature.

The other **15%** of corticospinal neurones, the **anterior corticospinal tract**, do not decussate and remain ipsilateral, eventually terminating in the upper thoracic spinal cord, and project bilaterally onto the motor neurones and interneurones that innervate the muscles of the upper trunk and neck.

The **extrapyramidal tract** neurones project to the spinal cord, where they synapse mainly onto interneurones. There are two groups: the **ventromedial** pathways, which terminate in the motor pools of the axial and proximal limb muscles, and the **dorsolateral** pathways, which terminate in the motor pools of the distal limb muscles. The **ventromedial pathways** comprise the **vestibulospinal tract**, which receives neurones from the vestibular system and is involved in the reflex control of balance, the **tectospinal tract**, which is involved in the coordination of eye and body movements, and the **reticulospinal tract**, which is concerned with regulating the excitability of extensor muscle reflexes. The **dorsolateral pathways** comprise mainly the **rubrospinal tract**, which originates in the **red nucleus** in the **midbrain** and projects to similar motor neurone pools as those served by the corticospinal tracts, and are involved with the **reflex control** of **flexor muscles**.

The cerebellum

The cerebellum is anatomically distinct from the rest of the brain and is connected to the brain stem by thick strands of afferent and efferent fibres through **three (cerebellar) peduncles**. Its primary function is the coordination and learning of movements, and it is made up of three functional and anatomical structures: the **spinocerebellum**, which is involved in the control of musculature and posture; the **cerebrocerebellum**, which is involved in the coordination and planning of limb movement; and the **vestibulocerebellum**, which is involved with posture and the control of eye movements. The **spinocerebellum** receives both sensory inputs from the spinal cord and motor inputs from the cerebral cortex. It regulates ongoing movements of axial and distal muscles, by comparison of the descending inputs with the ascending sensory feedback, and regulates muscle tone. The **cerebrocerebellum** receives inputs from the cerebral cortex, particularly the premotor cortex, and is primarily involved in the planning and initiation of movements, particularly involving the visual system. The **vestibulocerebellum** receives inputs and sends outputs to the vestibular nuclei in the medulla, and is involved in the regulation of balance, posture and the control of eye movements.

The cerebellum functions by acting as a **comparator**, comparing sensory and motor inputs and achieving coordinated movements that are both smooth and accurate. It can also function as a **timing device** in which it converts descending motor signals into a sequence of coordinated and smooth events. Finally, it can **store motor information** and regularly update it; therefore, given the right sequence of events, it can lead to the initiation of accurate learnt movements.

60 Proprioception and reflexes

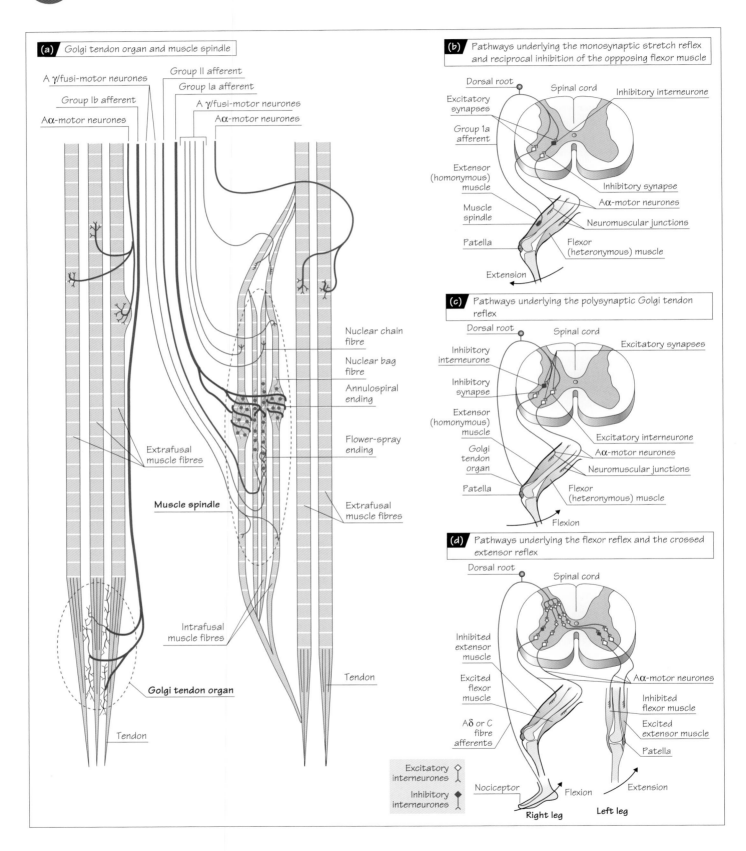

(a) Golgi tendon organ and muscle spindle

A γ/fusi-motor neurones
Group Ib afferent
Aα-motor neurones
Group II afferent
Group Ia afferent
A γ/fusi-motor neurones
Aα-motor neurones

Nuclear chain fibre
Nuclear bag fibre
Annulospiral ending
Flower-spray ending
Extrafusal muscle fibres
Extrafusal muscle fibres
Muscle spindle
Intrafusal muscle fibres
Golgi tendon organ
Tendon
Tendon

(b) Pathways underlying the monosynaptic stretch reflex and reciprocal inhibition of the opposing flexor muscle

Dorsal root
Spinal cord
Inhibitory interneurone
Excitatory synapses
Group 1a afferent
Inhibitory synapse
Aα-motor neurones
Neuromuscular junctions
Extensor (homonymous) muscle
Muscle spindle
Patella
Flexor (heteronymous) muscle
Extension

(c) Pathways underlying the polysynaptic Golgi tendon reflex

Dorsal root
Spinal cord
Excitatory synapses
Inhibitory interneurone
Inhibitory synapse
Extensor (homonymous) muscle
Golgi tendon organ
Patella
Excitatory interneurone
Aα-motor neurones
Neuromuscular junctions
Flexor (heteronymous) muscle
Flexion

(d) Pathways underlying the flexor reflex and the crossed extensor reflex

Dorsal root
Spinal cord
Inhibited extensor muscle
Excited flexor muscle
Aδ or C fibre afferents
Aα-motor neurones
Inhibited flexor muscle
Excited extensor muscle
Patella
Nociceptor
Flexion
Extension
Right leg
Left leg
Excitatory interneurones
Inhibitory interneurones

Physiology at a Glance, Third Edition. Jeremy P.T. Ward and Roger W.A. Linden.

We are aware of the orientation of our limbs with respect to one another, we can perceive the movements of our joints and we can accurately assess the amount of resistance (force) that opposes the movements we make. This ability is called **proprioception**. The three qualities of this modality are **position**, **movement** and **force**. The receptors or **proprioceptors** that mediate this modality are principally found in the joint capsules (**joint receptors**), muscles (**muscle spindles**) and tendons (**Golgi tendon organs**).

The **joint capsule** is compressed or stretched when the joint moves, and **mechanoreceptors** within it signal the position of the joint, as well as the direction and velocity of the movement. Individual receptors respond to the position of the joint, as well as the direction and the velocity of the movement, but not the force. The receptor types found in the joint capsule are **Ruffini-type** (slowly adapting) stretch receptors (Chapter 55).

Each muscle contains a number of small muscle fibres (**intrafusal muscle fibres**: 15–30 μm in diameter and 4–7 mm in length) that are thinner and shorter than the ordinary muscle fibre (**extrafusal muscle fibres**: 50–100 μm in diameter and varying in length from a few millimetres to many centimetres). Several intrafusal fibres are grouped together and encased in a connective tissue capsule, called the **muscle spindle**, a specialized receptor that responds to the stretch of a muscle (Fig. 60a). Muscle spindles lie in **parallel** to the extrafusal muscle fibres and are elongated when the muscle is stretched. The primary sensory innervation of the muscle spindle consists of afferent fibres which wind themselves around the centre of the intrafusal muscle fibres (**annulospiral ending**). These are large myelinated fibres (group Ia afferents). These endings are called **primary sensory endings** and, when excited, they evoke a **monosynaptic stretch reflex** involving an excitation of the **homonymous** α-motor neurones and reciprocal inhibition of the **heteronymous** α-motor neurones (Fig. 60b).

Many muscle spindles also have a **secondary sensory innervation** (group II afferent). They are thinner and end in **flower-spray endings** (they are not involved in the monosynaptic stretch reflex).

The intrafusal muscle fibres possess a motor innervation, the Aγ- or fusi-motor neurones. They are smaller in diameter than those innervating the extrafusal muscle fibres, the Aα-motor neurones.

Stretching the muscle, thereby extending both the extrafusal and intrafusal muscle fibres, can excite the muscle spindle. However, there is a second way to excite the primary muscle spindle ending – by contraction of the intrafusal muscle fibre brought about by excitation of the fusi- or γ-motor neurone. This does not change the overall length or tension of the entire muscle, as it is too weak; however, it is sufficient to stretch the central portion of the intrafusal fibres, inducing excitation in the primary sensory ending.

Contraction of the extrafusal fibres can be triggered or at least facilitated by the muscle spindle, either by stretch of the whole muscle or by activation of the fusi- or γ-motor neurones. They can complement each other or have a mutually cancelling effect. The threshold of the stretch reflex can also be varied by intrafusal activity.

The **Golgi tendon organs** are stretch receptors found in the muscle tendons (Fig. 60a). Each receptor is associated with the tendon fascicle of about 10 extrafusal muscle fibres, is surrounded by a capsule of connective tissue and is innervated by large myelinated afferent fibres (group Ib fibres). They are in **series** with the extrafusal muscle fibres and respond to tension in the muscle. They can respond both when the muscle contracts and when the muscle is stretched, unlike the muscle spindle which responds predominantly during stretching of the muscle. Golgi tendon organs protect against overloading and provide a protective reflex. In a functional sense, the segmental connections of the Ib fibres are a mirror image of the Ia fibres. A marked increase in muscle tension, whether resulting from stretch, contraction or a combination of the two, will result in **inhibitory** connections with the **homonymous** motor neurones and **excitatory** connections with the **heteronymous** motor neurones. However, none of these connections is monosynaptic; all involve at least two synapses (Fig. 60c).

The properties of the joint receptors make it very likely that they are primarily responsible for mediating the sense of position and movement. The most likely detectors of force sensation are the muscle spindles and Golgi tendon organs. Other receptors also contribute to the sense of force, as well as movement and position, such as mechanoreceptors in the skin.

Polysynaptic motor reflexes. Many receptors in the body, other than those found in muscle, can trigger motor reflexes. Experiments on animals, in which the spinal cord has been severed, have shown that many of these reflexes are restricted to the spinal cord. They are always polysynaptic. The most prominent examples of these reflexes are the flexor reflex and the crossed extensor reflex (Fig. 60d). If a pinprick is made to a toe, the stimulated limb is pulled away. The **flexor reflex** is a flexion of the knee and hip joints. It is a protective reflex, pulling the limb from the site of the noxious stimulus. The delay and the magnitude of the response are very much dependent on the stimulus intensity. The higher the intensity, the shorter the latency and the quicker the response. It can also be observed that the flexion of one limb is always accompanied by the extension of the limb on the other side of the body. In other words, there is an ipsilateral flexor reflex and a contralateral extensor reflex. This contralateral extensor reflex is called the **crossed extensor reflex**. These reflexes are not only enhanced in spinal animals, but also in newborn and premature babies, as during the days just after birth the higher levels of the brain are not fully developed.

Self-assessment MCQs

Chapter 1: Homeostasis and the physiology of proteins

1.1 Homeostasis
  (a) must always be restored using negative feedback mechanisms
  (b) provides for the tight regulation of all physiological variables
  (c) is the sum of all chemical reactions in the body
  (d) is a combination of positive and negative feedback mechanisms

1.2 A negative feedback mechanism comprises
  (a) detectors, comparators, a fixed set point and effectors
  (b) detectors, amplifiers, comparators, a set point and effectors
  (c) detectors, attenuators, comparators, a set point and effectors
  (d) detectors, comparators, a variable set point and effectors

1.3 Positive feedback
  (a) does not exist in physiological systems
  (b) is seen in the initiation of an action potential when sodium entry causes depolarization which in turn leads to potassium exit
  (c) is unstable and requires some mechanism to break the feedback loop
  (d) is switched off by negative feedback mechanisms

1.4 The primary structure of a protein
  (a) is determined by the sequence of amino acids
  (b) is a result of folding in the amino acid chain
  (c) is overseen by molecular chaperones such as heat shock proteins
  (d) has no effect on the functionality of the final molecule

Chapter 2: Body water compartments and physiological fluids

2.1 The percentages of water in each of the fluid compartments are approximately _____ intracellular fluid,_____ interstitial fluid and _____ plasma
  (a) 22%, 13%, 65%
  (b) 70%, 20%, 10%
  (c) 65%, 22%, 13%
  (d) 20%, 10%, 70%

2.2 Osmosis is
  (a) the active movement of water across a permeable membrane from a region of high solute concentration to that of a lower solute concentration
  (b) the passive movement of water across a semi-permeable membrane from a region of high solute concentration to that of a lower solute concentration
  (c) the active movement of water across a permeable membrane from a region of low solute concentration to that of a higher solute concentration
  (d) the passive movement of water across a semi-permeable membrane from a region of low solute concentration to that of a higher solute concentration

2.3 The colloidal osmotic or oncotic pressure is
  (a) important for controlling cell volume
  (b) a key determinant for fluid movement across capillary walls
  (c) largely determined by the Na^+ concentration in plasma
  (d) higher in interstitial fluid than in plasma

2.4 If a hypotonic fluid is ingested
  (a) intracellular fluid is diluted and cells swell
  (b) intracellular fluid is diluted but cells stay the same size
  (c) intracellular fluid is unaffected
  (d) intracellular fluid is concentrated and cells shrink

Chapter 3: Cells, membranes and organelles

3.1 Most cell membranes are composed principally of
  (a) DNA and ATP
  (b) proteins and lipids
  (c) chitin and starch
  (d) nucleotides and amino acids

3.2 Which of the following correctly matches the organelle with its function?
  (a) mitochondrion – anaerobic respiration
  (b) nucleus – protein assembly
  (c) Golgi apparatus – glycosylation
  (d) lysosome – movement

3.3 Aerobic resynthesis of ATP occurs in the
  (a) mitochondria
  (b) cytosol
  (c) lipid membrane
  (d) sarcoplasmic reticulum

3.4 G-protein-coupled receptors
  (a) can activate signal proteins called G-proteins
  (b) possess five membrane spanning segments
  (c) can detect signal molecules, such as neurotransmitters or hormones, in the intracellular fluid
  (d) penetrate the entire thickness of the phospholipids bilayer, with the intramembrane segments made up of chains of hydrophilic amino acid residues

Chapter 4: Membrane transport proteins and ion channels

4.1 The Na^+–K^+ ATPase (sodium pump) is
  (a) a symporter
  (b) an uniporter
  (c) an antiporter
  (d) an example of secondary active transport

4.2 Which of the following is characterized by carrier-mediated transport down a chemical concentration gradient?
  (a) active transport
  (b) facilitated diffusion
  (c) diffusion
  (d) osmosis

4.3 Which of the following statements is **not** true about ligand-gated ion channels?
  (a) Neurotransmitters can act as external triggers to open ligand-gated ion channels

Physiology at a Glance, Third Edition. Jeremy P.T. Ward and Roger W.A. Linden.
 © 2013 John Wiley & Sons, Ltd. Published 2013 by John Wiley & Sons, Ltd.

(b) Ligand-gated ion channels are found in the cell membrane

(c) Ligand-gated ion channels can be activated by cyclic AMP

(d) Differences in membrane potential can affect whether ligand-gated ion channel receptors open or close

4.4 Ion channels

(a) are always selective for a particular ion

(b) are opened and closed by a process called gating which is brought about by changes in the conformation of protein subunits

(c) cannot be gated by pH changes in the external environment

(d) provide a charged, hydrophobic pore through which ions can diffuse across the lipid bilayer

Chapter 5: Biological electricity

5.1 If the extracellular potassium concentration rises from 4 mmol/L to 5 mmol/L, what happens to the resting membrane potential?

(a) it becomes more negative

(b) it becomes less negative

(c) it remains the same

(d) it initiates an action potential

5.2 In the resting membrane potential state

(a) there are more Na^+ than K^+ channels in an open state

(b) the permeability to K^+ ions exceeds that of Na^+ ions

(c) the K^+:Na^+ permeability ratios are the same in both excitable and non-excitable cells

(d) the exact potential can be calculated using the Nernst equation

5.3 During the rising phase of an action potential

(a) voltage-gated Na^+ channels open

(b) voltage-gated K^+ channels open

(c) voltage-gated Na^+ channels close

(d) voltage-gated K^+ channels close

5.4 For an action potential to occur

(a) the generator potential must reach or exceed threshold

(b) the membrane potential must be out of the relative refractory period

(c) Na^+ efflux must exceed K^+ influx

(d) all the above

5.5 Which is the correct sequence of events that follows a threshold potential?

1 the membrane becomes depolarized

2 Na^+ channels open and Na^+ ions diffuse inward

3 the membrane becomes repolarized

4 K^+ channels open and K^+ ions diffuse outward while Na^+ is actively transported out of the cell

(a) 3, 2, 4 and 1

(b) 2, 1, 4 and 3

(c) 2, 1, 3 and 4

(d) 1, 2, 4 and 3

Chapter 6: Conduction of action potentials

6.1 Which of the following is true about the conduction of action potentials?

(a) thicker axons conduct action potentials faster because there is a faster current flow in thicker axons

(b) action potentials travel faster in unmyelinated nerve fibres than myelinated nerve fibres

(c) the action potential in myelinated axons is generated only at the nodes of Ranvier

(d) the action potential in an unmyelinated nerve has an amplitude that declines with distance from its site of origin

6.2 Saltatory conduction

(a) can be up to 100 times faster than the conduction along the fastest unmyelinated axon

(b) occurs because the cell membrane beneath the myelin sheath has a higher density of Na^+ channels

(c) is where depolarization jumps from one node of Ranvier to another and by doing so uses up more energy than in unmyelinated conduction

(d) can occur both orthodromically and antidromically if the axon is stimulated in its middle

6.3 Nerve fibres that conduct at between 70 and 120 m/s are

(a) motor to muscle spindles

(b) afferents from primary muscle spindles and motor to skeletal muscles

(c) large unmyelinated fibres

(d) afferents from Golgi tendon organs

6.4 The compound action potential

(a) obeys the all-or-nothing law

(b) is usually a multi-peaked response with the slowest conducting fibres seen in the first peak

(c) is usually a multi-peaked response with the fastest conducting fibres seen in the first peak

(d) can be recorded by using intracellular recording electrodes

Chapter 7: The autonomic nervous system

7.1 The autonomic nervous system

(a) controls smooth muscle, skeletal muscle and cardiac muscle

(b) controls smooth muscle, cardiac muscle and glandular activity

(c) mainly provides the afferent pathways for involuntary control of organs

(d) is subdivided into parasympathetic and sympathetic systems that mostly work antagonistically on blood vessels

7.2 Sympathetic neurones

(a) originate in the ventral horn of the spinal cord

(b) release adrenaline (epinephrine) at their terminals

(c) can release acetylcholine at their terminals

(d) have peripheral ganglia found close to or in the target organ

7.3 Parasympathetic neurones

(a) originate in the brain stem and then run via the vagus to the paravertebral ganglia

(b) release noradrenaline (norepinephrine) at their terminals which in turn activates adrenergic receptors

(c) cause vasodilatation in the blood vessels of the external genitalia and salivary glands

(d) coordinate the so-called '*flight or fight*' response

7.4 In the synapse

(a) action potentials are usually transmitted from one neurone to another by electronic coupling

(b) transmitter substances are stored in vesicles that are about 200 nm in diameter

(c) the presynaptic axon terminates in a bulbous swelling called a bouton which is separated from the postsynaptic neurone by a synaptic cleft

(d) the arrival of an action potential at the nerve ending causes an influx of Na^+, which in turn causes the fusion of the vesicles containing the transmitter substance with the cell membrane and the subsequent release of the neurotransmitter

Chapter 8: Blood

8.1 Erythrocyte production

(a) occurs in the liver and spleen in neonates

(b) is impaired at high altitude

(c) is stimulated by hyperoxia

(d) is stimulated by erythropoietin released from the liver and spleen

8.2 White blood cells

(a) do not contain a nucleus

(b) have a life span of 100–130 days

(c) include lymphocytes which destroy bacteria by a process called phagocytosis

(d) can release mediators such as histamine that are involved in inflammatory processes

8.3 Reticulocytes

(a) normally comprise 1–2% of red cells in the blood

(b) are nucleated

(c) have a lifespan of 120 days

(d) are formed from normoblasts after leaving the bone marrow

8.4 Which one of the following is **not** associated with a plasma protein?

(a) copper transport

(b) thalassemia

(c) the immune system

(d) capillary filtration

Chapter 9: Platelets and haemostasis

9.1 Platelets

(a) release ADP after activation

(b) are also called megakaryocytes

(c) can only be activated via their glycoprotein receptors

(d) have dense vesicles containing histamine

9.2 During primary homeostasis

(a) fibrinogen is converted to fibrin

(b) the initial vasoconstriction does not require platelet activation

(c) thromboxane A_2 causes platelet adhesion

(d) activation of lipooxygenase is a vital step

9.3 Activation of thrombin

(a) is triggered by thrombomodulin

(b) always requires the presence of activated platelets

(c) is Ca^{2+} independent

(d) depends on factor X

9.4 Fibrin

(a) requires factor XIII to polymerise

(b) is destroyed by active protein C

(c) binds plasminogen

(d) activates von Willebrand factor

Chapter 10: Defence: Inflammation and immunity

10.1 Adaptive immunity is

(a) acquired and results in immunological memory

(b) specific and mediated by natural killer cells

(c) mediated by cytokines

(d) non-specific

10.2 Which of the following is **not** true regarding inflammation?

(a) characterized by pain, heat, redness and swelling

(b) it increases vascular permeability

(c) the changes are initiated by cytokines released by neutrophils

(d) migrating neutrophils play an important role

10.3 Which form of immunoglobulin is most likely to be involved in secretions?

(a) IgE

(b) IgG

(c) IgA

(d) IgM

10.4 T helper cells

(a) respond to antigens of any type

(b) only recognize antigens presented by major histocompatibility complex type 1

(c) phagocytose pathogens

(d) coordinate the immune response via cytokines

Chapter 11: Principles of diffusion and flow

11.1 Passive diffusion

(a) involves a carrier medium

(b) requires the expenditure of energy

(c) refers to movement down a concentration gradient

(d) is described by Darcy's law

11.2 The rate of diffusion in a solution is described by

(a) Fick's law

(b) Darcy's law

(c) Poiseuille's law

(d) Laplace's law

11.3 The flow through the majority of the cardiovascular system at rest

(a) is laminar

(b) is turbulent

(c) is described by Poiseuille's law which states that flow is dependent on the pressure difference across the ends of a tube and the resistance provided by the tube

(d) is not affected by changes in the viscosity of blood

11.4 If a small alveolus was connected to a large alveolus in the lungs in the absence of surfactant

(a) the large alveolus would empty into the small alveolus

(b) nothing would change

(c) the small alveolus would empty into the large alveolus

(d) both alveoli would empty

Chapter 12: Skeletal muscle and its contraction

12.1 The two principal contractile proteins in skeletal muscles are

(a) actin and troponin

(b) actin and tropomyosin

(c) actin and myosin

(d) myosin and tropomyosin

12.2 The trigger to initiate the contractile process in skeletal muscle is
  (a) Ca^{2+} binding to tropomyosin
  (b) ATP binding to the myosin cross-bridges
  (c) Ca^{2+} binding to troponin
  (d) K^+ binding to myosin

12.3 After Ca^{2+} ions are released from the sarcoplasmic reticulum they
  (a) initiate an action potential
  (b) bind to actin
  (c) cause sodium channels to open in the sarcolemmal membrane
  (d) bind to troponin

12.4 The A-bands of skeletal muscle do not change their width during muscle contraction because
  (a) the thick filaments are not involved in the sliding filament theory
  (b) the A-bands are the thick filaments themselves which do not shorten
  (c) the A-bands extend beyond the sarcomere
  (d) the A-bands are the thin filaments themselves which do shorten

Chapter 13: Neuromuscular junction and whole muscle contraction

13.1 At the neuromuscular junction
  (a) the muscle membrane possesses muscarinic receptors
  (b) there is a one-to-one transmission of excitatory impulses from the motor neurone to the muscle fibres it innervates
  (c) the motor nerve endings secrete noradrenaline (norepinephrine)
  (d) the typical summed end plate potential (EPP) is usually 10 times the potential necessary to trigger an action potential

13.2 During isotonic contraction
  (a) the A-band shortens
  (b) the sarcomere and the I-band shorten
  (c) the A-band and the I-band shorten
  (d) the sarcomere and the A-band shorten

13.3 An action potential in the motor end plate rapidly spreads to the central portions of the muscle cells by means of the
  (a) Z-lines
  (b) sarcoplasmic reticulum
  (c) pore in the plasma membrane
  (d) transverse tubules

13.4 During isometric contraction of skeletal muscle
  (a) the I-bands shorten and the A-bands stay the same length
  (b) the thick and thin filaments slide past each other
  (c) the sarcomere length does not change
  (d) none of the above

Chapter 14: Motor units, recruitment and summation

14.1 'Motor unit' refers to
  (a) a single motor neurone and all the muscle fibres it innervates
  (b) a single muscle fibre and all the motor neurones that innervate it

  (c) all the motor neurones supplying a single muscle
  (d) a pair of antagonist muscles

14.2 Which of the following is **not** true regarding the comparison of type I (slow oxidative) and type IIB (fast glycolytic) skeletal muscle fibres?
  (a) type I fibres have a richer capillary supply
  (b) type I fibres have smaller cell bodies
  (c) type I fibres fatigue more readily
  (d) type I fibres have a smaller diameter

14.3 The sustained contraction by skeletal muscles in which individual twitches cannot be detected is called
  (a) summation
  (b) recruitment
  (c) tetanus
  (d) twitching

14.4 All of the following are involved in producing graded contractions of an entire muscle **except**
  (a) small motor units (those with fewer muscle fibres, e.g. 5) are utilized for fine movements whilst large motor units (those with more muscle fibres, e.g. 600) are utilized for gross movements
  (b) discharging many motor units at one time (spatial recruitment) allows increased force to be generated
  (c) discharging individual motor units at a more rapid rate (temporal recruitment) allows increased force to be generated
  (d) contraction of only a portion of the muscle fibres within a single motor unit during a finely graded movement

Chapter 15: Cardiac and smooth muscle

15.1 Myocytes are
  (a) approximately $100\,\mu m \times 50\,\mu m$ in dimensions
  (b) multinucleated
  (c) linked to one another at their ends by electronic junctions called intercalated discs
  (d) low in mitochondria

15.2 Regarding smooth muscle
  (a) it has actin and myosin filaments organized into myofibrils
  (b) it is multinucleated
  (c) like cardiac muscle, the cells are joined together by gap junctions called intercalated discs
  (d) it is capable of prolonged contraction without fatigue and with little energy consumption

15.3 Which of the following statements is **not** true?
  (a) smooth muscle has no troponin
  (b) smooth muscle is innervated by neurones of the autonomic nervous system
  (c) multiunit smooth muscle is made up of individual fibres not connected by gap junctions
  (d) smooth muscle cells are not affected by local tissue factors and hormones

15.4 The effect of nerve stimulation on smooth muscle is
  (a) excitatory only
  (b) inhibitory only
  (c) both excitatory or inhibitory
  (d) modulatory only

Chapter 16: Introduction to the cardiovascular system

16.1 The cardiovascular system
(a) contains approximately 7 L of blood in a 70 kg man
(b) is arranged mostly in series with each tissue receiving blood from the aorta
(c) is arranged mostly in parallel with each tissue receiving blood from the aorta
(d) maintains a constant pressure throughout the system

16.2 Smooth muscle is not to be found in the walls of
(a) veins
(b) venules
(c) arterioles
(d) capillaries

16.3 In the heart, the cardiac output is approximately_____/min at rest, rising to above_____/min during exercise. The stroke volume at rest is approximately_____.
(a) 20 L, 100 L, 1 L
(b) 5 L, 100 L, 70 mL
(c) 500 mL, 20 L, 7 mL
(d) 5 L, 20 L, 70 mL

16.4 The mean arterial blood pressure is
(a) 110 mmHg in a healthy person
(b) calculated by averaging the systolic blood pressure and the diastolic blood pressure
(c) estimated as the diastolic blood pressure plus one third of the systolic blood pressure
(d) estimated as the diastolic blood pressure plus one third of the pulse pressure

Chapter 17: The heart

17.1 Blood flows from the right atrium into the right ventricle via
(a) the mitral valve
(b) the semilunar valves
(c) the tricuspid valve
(d) the AV node

17.2 Conduction of impulses between the atria and the ventricles is channelled through the
(a) AV node
(b) annulus fibrosus
(c) SA node
(d) bundle of His

17.3 The atrioventricular node
(a) conducts impulses rapidly allowing almost immediate activation of ventricular muscle
(b) delays impulses for about 200 ms, allowing time for atrial contraction to complete ventricular filling
(c) is connected to specialized wide, fast conducting myocytes in the bundle of His and Purkinje fibres
(d) is a band of fibrous connective tissue which separates the atria from the ventricles

17.4 Which of the following statements concerning coronary circulation is true?
(a) Left ventricular perfusion only occurs during diastole
(b) The heart receives a rich blood supply from the left and right coronary arteries arising from the coronary sinus
(c) The coronary veins run parallel to the right coronary arteries and empty into the aortic sinus
(d) During systole, contraction of the ventricles compresses the coronary arteries and suppresses blood flow

Chapter 18: The cardiac cycle

18.1 The aortic valve
(a) prevents the backflow of blood into the aorta during ventricular diastole
(b) prevents the backflow of blood into the left ventricle during ventricular diastole
(c) prevents the backflow of blood into the left ventricle during ventricular systole
(d) prevents the backflow of blood into the aorta during ventricular systole

18.2 Which of the following indicates the causes of the first and second heart sound in the correct order?
(a) atrial systole – ventricular systole
(b) semilunar valve closure – atrioventricular valve closure
(c) ventricular diastole – semilunar valve closure
(d) ventricular systole – ventricular diastole

18.3 During which phases of the cardiac cycle do the atrioventricular valves remain open?
(a) atrial diastole
(b) isovolumetric ventricular relaxation
(c) isovolumetric ventricular contraction
(d) passive filling

18.4 The ventricular end-diastolic volume in man is approximately _____ and the end-diastolic pressure is_____
(a) 130 mL, >100 mmHg
(b) 130 mL, <10 mmHg
(c) 70 mL, <10 mmHg
(d) 70 mL, >100 mmHg

Chapter 19: Initiation of the heart beat and excitation–contraction coupling

19.1 Action potentials in ventricular muscles
(a) are identical to those in skeletal muscles except for the duration of the action potential
(b) have a plateau phase caused by the delay in the opening of K^+ channels
(c) are initiated when ventricular myocytes are depolarized to a threshold potential of −50 mV
(d) have a refractory period which prevents another action potential from being initiated until the muscle relaxes

19.2 The cells of the sinoatrial node
(a) have a resting potential of −90 mV
(b) have action potentials which exhibit a slow upstroke because of the presence of L-type calcium channels
(c) are the only cells in the heart that can act as pacemaker cells
(d) are directly affected by noradrenaline (norepinephrine) and acetylcholine in that they slow down and speed up the heart respectively

19.3 Excitation–contraction coupling in cardiac ventricular cells requires
(a) efflux of Na^+ ions
(b) efflux of K^+ ions
(c) influx of Ca^{2+} ions
(d) influx of Cl^- ions

19.4 Noradrenaline (norepinephrine)
  (a) has a positive inotropic effect on the heart muscle cells whilst acetylcholine has a negative inotropic effect
  (b) is released by the sympathetic fibres innervating the SA node only
  (c) has a positive chronotropic effect on cardiac cells by increasing the rate of decay of the pacemaker potential
  (d) slows down the Ca^{2+} sequestration into the sarcoplasmic reticulum, thereby increasing contractility of the cardiac cells

Chapter 20: Control of cardiac output and Starling's law of the heart

20.1 Cardiac output is
  (a) the volume of blood pumped per minute by both ventricles
  (b) the volume of blood flowing through the systemic circulation per minute
  (c) not affected by the filling pressure (preload)
  (d) approximately 5 L/min at rest rising to over 100 L/min during exercise

20.2 Starling's law of the heart
  (a) states that 'the stroke volumes of the left and right ventricles are matched'
  (b) concerns the relationship between the degree of stretch of cardiac muscle and the force of contraction
  (c) causes an increase of contractility of cardiac muscle
  (d) can equally be applied to skeletal as well as cardiac muscle

20.3 Constriction of the veins
  (a) decreases venous compliance and therefore increases central venous pressure (CVP) (b) increases venous resistance and therefore decreases CVP
  (c) increases the slope of the vascular function curve (d) reduces venous return

20.4 Which of the following statements is **not** true?
  (a) The mechanism underlying Starling's law of the heart is intrinsic to cardiac muscle
  (b) The autonomic nervous system provides an important extrinsic influence on cardiac output
  (c) Sympathetic stimulation can cause an increase in contractility, which can be an increase in force without a change in length
  (d) Cardiac output is only affected by the filling pressures of the left side of the heart and the effects of the autonomic nervous system on the heart rate and contractility

Chapter 21: Blood vessels

21.1 Smooth muscle and elastin filaments are found in the
  (a) tunica intima
  (b) tunica adventitia
  (c) tunica media
  (d) elastic lamina

21.2 Fenestrated capillaries
  (a) are more permeable than continuous capillaries and are found in the skin and muscle
  (b) are less permeable than continuous capillaries and are found in endocrine glands and intestinal villa
  (c) have fewer tight junctions than continuous capillaries and have pores in their endothelial cells
  (d) are found in bone marrow, liver and spleen

21.3 The following substances are all vasodilators
  (a) angiotensin II, histamine and bradykinin
  (b) histamine, bradykinin and substance P
  (c) acetylcholine, bradykinin and endothelin-1
  (d) substance P, noradrenaline (norepinephrine) and phospholipase C

21.4 The endothelium
  (a) can synthesize a number of important vasoconstrictors such as nitric oxide (NO)
  (b) can synthesize a number of important vasodilators such as endothelium-derived relaxing factor (EDRF)
  (c) can only synthesize vasodilators
  (d) plays a minor role in the regulation of vascular tone

Chapter 22: Control of blood pressure and blood volume

22.1 An increase in the mean arterial blood pressure may result from
  (a) an increase in peripheral resistance
  (b) an increase in heart rate
  (c) an increase in venous return
  (d) a, b and c

22.2 Baroreceptors
  (a) in the carotid and aortic bodies, send impulses via the glossopharyngeal and vagus nerves to the medulla of the brain stem
  (b) are sensors for mean arterial blood pressure that respond to stretching of the carotid sinus and aortic arch
  (c) do not exhibit adaptation
  (d) are most sensitive between 120 and 180 mmHg

22.3 A fall in mean arterial blood pressure
  (a) causes an increase in baroreceptor activity and a reduction in renal perfusion pressure
  (b) along with sympathetic stimulation will activate the renin–angiotensin system and the production of angiotensin II
  (c) increases Na^+ and water excretion
  (d) is sensed by atrial baroreceptors

22.4 Which of the following statements is **not** true?
  (a) 20% of the blood volume can be lost without significant problems
  (b) An acute fall in mean arterial blood pressure can result from a sudden blood loss, a profound vasodilatation or an acute failure of the heart to pump blood
  (c) Following the loss of less than 20% of the blood volume, the volume of blood is restored within minutes of the blood loss due to arteriolar constriction
  (d) A loss of 30–50% of blood volume can be survived if a blood transfusion is given within an hour

Chapter 23: The microcirculation, filtration and lymphatics

23.1 Which of the following substances are highly lipophilic and therefore can cross the endothelial lipid bilayer membrane easily?
  (a) oxygen and carbon dioxide
  (b) water
  (c) glucose and Na^+ ions
  (d) plasma proteins

23.2 In a capillary, the net flow of water between the blood and the interstitial fluid can change its direction between the arterial and venous ends of the capillary. This difference at the two ends reflects a change in which of the following factors?
  (a) osmotic pressure
  (b) vessel diameter
  (c) blood velocity
  (d) blood pressure

23.3 Lymph fluid is returned into the general circulation via the
  (a) left atrium
  (b) superior vena cava
  (c) subclavian vein
  (d) azygous vein

23.4 Which of the following statement is **not** true?
  (a) oedema is the swelling of the tissues due to excess fluid in the interstitial space
  (b) oedema can be caused by an increased venous pressure
  (c) the colloidal osmotic pressure normally varies between ~25 mmHg at the arteriolar end of the capillary and ~15 mmHg at the venous end
  (d) lymphatic capillaries are blind-ended bulbous tubes approximately 15–75 μm in diameter.

Chapter 24: Local control of blood flow and special circulations

24.1 Autoregulation is the ability of a tissue to maintain a constant blood flow in the face of variations of pressure. It is particularly important in the
  (a) brain, kidneys and heart
  (b) skin and gut
  (c) skeletal muscle
  (d) lungs

24.2 The following substances are autocoids (i.e. local hormones)
  (a) bradykinin, nitric oxide and histamine
  (b) serotonin, histamine and adenosine
  (c) bradykinin, histamine and serotonin
  (d) nitric oxide, adenosine and CO_2

24.3 Which of the following statements is true?
  (a) skeletal muscle comprises approximately 20% of body weight yet in exercise can take over 80% of cardiac output
  (b) in skeletal muscles at rest, the majority of capillaries are not perfused, as their arterioles are constricted
  (c) in skeletal muscle, capillaries are recruited during exercise by metabolic hyperaemia caused by the release of CO_2 and Ca^{2+}
  (d) the endothelial cells of the capillaries in the brain have very few tight junctions so there is free movement of fluids between the blood and the cerebrospinal fluid

24.4 Blood flow
  (a) through the skin is controlled by both sympathetic and parasympathetic nerve fibres
  (b) through skeletal muscle involves arteriovenous anastomoses which directly link arterioles and venules, by-passing the capillaries
  (c) in the pulmonary circulation is not controlled by autonomic nerves or metabolic products
  (d) in the pulmonary circulation is increased during hypoxia due to vasodilatation in small arteries

Chapter 25: Introduction to the respiratory system

25.1 The U-shaped rings that form the framework of the trachea and help to keep it open are composed of
  (a) skeletal muscle
  (b) bone
  (c) cartilage
  (d) fibroelastic tissue

25.2 Alveolar type II pneumocytes are
  (a) cuboidal cells that secrete surfactant
  (b) squamous cells involved in gas exchange
  (c) ciliated cells that move the mucus
  (d) columnar cells that secrete mucus

25.3 Which muscles are involved in quiet inspiration?
  (a) the internal and external intercostal muscles
  (b) the diaphragm
  (c) the abdominal muscles
  (d) the diaphragm and external intercostal muscles

25.4 In order for the lungs to function normally the intrapleural pressure must
  (a) alternate between being less than or greater than atmospheric pressure
  (b) be the same as atmospheric pressure
  (c) be between 0.2 and 0.5 kPa
  (d) be between −0.2 and −0.5 kPa

Chapter 26: Lung mechanics

26.1 Which of the following statements is true?
  (a) the change in the volume of the lungs with pressure is a measure of their compliance
  (b) the total compliance of the chest is determined solely by the compliance of the lungs
  (c) the recoil of the lungs assists inspiration
  (d) pulmonary surfactant maintains a constant low surface tension in the alveoli

26.2 Which of the following statements is/are true?
  1 neonatal respiratory distress syndrome is caused by lack of sufficient surfactant
  2 surfactant is produced by type I pneumocytes
  3 surfactant creates hysteresis
  4 the pressure due to surface tension is greater in a soap bubble of greater radius
  (a) 1 only
  (b) 3 only
  (c) 2 and 4
  (d) 1, 3 and 4

26.3 Dynamic compression of airways
  (a) is caused by bronchoconstrictors such as histamine, prostaglandins and leukotrienes
  (b) occurs during forced expiration
  (c) is brought about by reflex activation of parasympathetic nerves
  (d) does not occur in normal expiration because the intrapleural pressure remains positive throughout

26.4 Peak expiratory flow rate (PEFR)
  (a) is not dependent on initial lung volume
  (b) provides as much information as the forced expiratory volume against time
  (c) increases as the airways resistance increases

(d) is useful when following the progression and treatment of a disease such as asthma

Chapter 27: Transport of gases and the gas laws

27.1 According to Dalton's law the partial pressure of O_2 (Po_2) in dry air at a barometric pressure of $101\,kPa$ (sea level) is____ kPa and at a barometric pressure of $34\,kPa$ (summit of Everest) is____ kPa
  (a) 21.2, 7.14
  (b) 21.0, 8.0
  (c) 19.9, 5.8
  (d) 20, 6.15

27.2 Typical partial pressures for alveolar air at $37\,^\circ C$ and 100% humidity are Pco_2 ____ kPa, Po_2 ____kPa, P_{N2} ____ kPa and saturated water vapour pressure ____ kPa
  (a) 0, 19.9, 74.8, 6.3
  (b) 0.05, 0.13, 0.75, 0.06
  (c) 5.3, 13.3, 76.1, 6.3
  (d) 0, 21.2, 79.8, 0

27.3 The quantity of gas dissolving in a fluid is described by
  (a) Dalton's law
  (b) Boyle's law
  (c) Charles' law
  (d) Henry's law

27.4 CO_2 crosses the alveolar–capillary membrane
  (a) faster than O_2 because it has a greater molecular weight
  (b) slower than O_2 because it has a greater molecular weight
  (c) faster than O_2 because it is more soluble in biological membranes
  (d) slower than O_2 because it is less soluble in biological membranes

Chapter 28: Carriage of oxygen and carbon dioxide by the blood

28.1 The O_2–haemoglobin dissociation curve is shifted to the right by
  (a) a decrease in acidity
  (b) an increase in Pco_2
  (c) a decrease in temperature
  (d) a decrease in 2,3-DPG

28.2 CO_2 is transported in the blood approximately
  (a) 60% as bicarbonate, 30% as carbamino compounds and 10% dissolved
  (b) 60% as bicarbonate, 20% as carbamino compounds and 20% dissolved
  (c) 50% as bicarbonate, 40% as carbamino compounds and 10% dissolved
  (d) 50% as bicarbonate, 35% as carbamino compounds and 15% dissolved

28.3 Which of the following statements is **not** true?
  (a) fetal haemoglobin binds 2,3-DPG less strongly than does adult haemoglobin, thus shifting the dissociation curve to the right
  (b) fetal haemoglobin facilitates the transfer of oxygen from maternal blood to the fetus
  (c) carbon monoxide binds 240 times more strongly than oxygen to haemoglobin
  (d) in anaemia, arterial Po_2 and O_2 saturation remain normal

28.4 Hyperventilation
  (a) is rapid breathing during exercise
  (b) reduces the arterial Pco_2 and increases arterial Po_2 substantially
  (c) can lead to light-headedness, visual disturbances and tetanus
  (d) can be caused by pain, hysteria and strong emotion

Chapter 29: Control of breathing

29.1 In the control of breathing
  (a) afferents from the stretch receptors feed into the pneumotaxic centre in the pons to modulate breathing
  (b) when expiratory neurones fire they inhibit the activity of inspiratory neurones
  (c) voluntary control of breathing is possible because motor neurones from the cerebral cortex directly innervate respiratory group neurones in the medulla, altering their activity
  (d) juxtapulmonary receptors detect fluid in the alveolar walls to cause an increased rate of breathing and an increase of both heart rate and blood pressure

29.2 Ventilation may be increased
  (a) by a decrease in plasma pH (acidosis)
  (b) by breathing O_2-enriched air
  (c) by stimulation of the intracranial chemoreceptors by hypoxia
  (d) by stimulation of the carotid sinus by hypercapnia

29.3 The receptors that are located on the alveolar and bronchial walls close to the capillaries are called
  (a) pulmonary stretch receptors
  (b) irritant receptors
  (c) proprioceptors
  (d) juxtapulmonary receptors

29.4 The central chemoreceptor
  (a) is a collection of neurones near the ventrolateral surface of the pons
  (b) responds to changes in both blood Pco_2 and Po_2
  (c) responds to a rise in H^+ in the CSF and leads to an increase in ventilation
  (d) is responsible for only 20% of the response to CO_2 in humans

Chapter 30: Ventilation–perfusion matching and right to left shunts

30.1 In a situation where there is a ventilation–perfusion mismatch with a V_A/Q value of less than unity, i.e. shunt effect
  (a) there is no gas exchange
  (b) the blood has a lower than normal Po_2, and a higher than normal Pco_2
  (c) the blood has a higher than normal Po_2, and a lower than normal Pco_2
  (d) there is excessive ventilation

30.2 Ventilation with O_2-enriched air will
  (a) improve oxygenation in regions of high V_A/Q
  (b) improve oxygenation in regions of low V_A/Q
  (c) improve oxygenation in shunts
  (d) have no effect of V_A/Q in any situation

30.3 Gravity

  (a) causes a higher blood flow at the base of the lungs than at the apex of the lungs

  (b) causes a less negative intrapleural pressure at the apex of the lungs than at the base of the lungs

  (c) causes ventilation to be greater at the apex of the lungs than at the base of the lungs

  (d) has no effect on V_A/Q

30.4 Anatomical right to left shunts where oxygenated blood from the lungs is diluted by venous blood

  (a) occur only in disease or congenital heart malformations

  (b) account for approximately 5% of cardiac output in healthy individuals

  (c) commonly result in a low arterial P_{O_2} and a high P_{CO_2}

  (d) can occur in diseases in which regions of the lungs are not ventilated

Chapter 31: Introduction to the renal system

31.1 Which one of these statements is false? The nephron

  (a) has endocrine functions

  (b) is under endocrine control

  (c) absorbs more ions and molecules than it secretes

  (d) lies exclusively within the renal cortex

31.2 In the kidney

  (a) 85% of the nephrons are juxtamedullary nephrons and 15% are cortical nephrons

  (b) the proximal, distal and collecting ducts are found in the medulla

  (c) the proximal tubule walls are formed of thin, flat squamous epithelial cells with no microvilli

  (d) urine is propelled through the ureter into the bladder by peristalsis

31.3 Which of these statements is true?

  (a) the kidney receives approximately 30% of cardiac output

  (b) the capillaries of the glomerulus are the site of filtration and drain into the efferent vein

  (c) the only blood supply to the medulla is the vasa recta

  (d) the blood flow through the cortex and the medulla are approximately equal

31.4 Micturition is

  (a) totally controlled by higher voluntary centres

  (b) initiated by a spinal reflex when urine pressure reaches a critical level

  (c) brought about by the contraction of skeletal muscle (the detrusor muscle) surrounding the bladder

  (d) totally controlled by the internal and external urethral sphincters

Chapter 32: Renal filtration

32.1 In the glomerulus, if the diameter of the afferent arteriole is smaller than the diameter of the efferent arteriole, then

  (a) the net filtration pressure will decrease

  (b) blood pressure in the glomerulus will increase

  (c) GFR will increase

  (d) there will be little change in GFR

32.2 Glomerular filtration rate is approximately _____ mm/min, whereas renal plasma flow is approximately _____ mL/min

  (a) 50, 300

  (b) 125, 600

  (c) 200, 300

  (d) 125, 250

32.3 Which of the following does **not** reduce glomerular filtration rate?

  (a) renal disease

  (b) local vasoconstrictors

  (c) sympathetic activation

  (d) angiotensin II

32.4 Glomerular filtration rate can be calculated using any substance that

  (a) is freely filtered by the nephron

  (b) is freely filtered and secreted by the nephron

  (c) is freely filtered and neither reabsorbed nor secreted by the nephron

  (d) is freely filtered and completely reabsorbed by the nephron

Chapter 33: Reabsorption and secretion and the proximal tubule

33.1 Reabsorption and secretion in the proximal tubule involves

  (a) diffusion through tight junctions and lateral intercellular spaces

  (b) active transport involving proteins called transporters

  (c) facilitated diffusion

  (d) all of the above

33.2 Which of the following is not reabsorbed in the proximal tubule?

  (a) glucose

  (b) amino acids

  (c) organic acids

  (d) bicarbonate

33.3 The tubular transport maximum (T_m) for glucose is approximately

  (a) 21 mg/min

  (b) 21 mmol/min

  (c) 11 mmol/min

  (d) 380 mmol/min

33.4 In the proximal tubule, water

  (a) is actively reabsorbed

  (b) is actively secreted

  (c) reabsorption increases tubular concentrations of Cl^-, K^+, Ca^{2+} and urea

  (d) content is decreased by about 50%

Chapter 34: The loop of Henle and the distal nephron

34.1 In which part of the nephron does antidiuretic hormone (ADH) have its main effect?

  (a) the proximal tubule

  (b) the descending limb of the loop of Henle

  (c) the ascending limb of the loop of Henle

  (d) the distal tubule and the collecting duct

34.2 Fluid entering the loop of Henle is normally _____ with plasma, and that entering the distal tubule is normally _____ with plasma.

  (a) isotonic, hypertonic

  (b) hypertonic, hypotonic

(c) isotonic, hypotonic

(d) hypotonic, hypertonic

34.3 Which of the following is **not** true of the counter-current multiplier system?

(a) the ascending limb of the loop of Henle transports Na^+ and Cl^- by active transport

(b) the descending limb of the loop of Henle is impermeable to water

(c) the fluid of the ascending limb of the loop of Henle becomes relatively dilute (low osmolarity)

(d) the vasa recta removes water reabsorbed from both the loop of Henle and the collecting ducts

34.4 At its most concentrated, final urine can be about _____ times more concentrated than plasma

(a) 2

(b) 8

(c) 5

(d) 10

Chapter 35: Regulation of plasma osmolality and fluid volume

35.1 Plasma osmolality is

(a) decreased in water deficiency

(b) increased by the ingestion of water

(c) controlled by osmoreceptors in the anterior pituitary

(d) regulated by antidiuretic hormone (ADH)

35.2 Antidiuretic hormone (ADH)

(a) is an nine-amino acid peptide formed from a precursor synthesized in the pituitary

(b) is stored in secretory granules in the anterior pituitary

(c) is released from secretory granules by action potentials from osmoreceptors

(d) causes vasoconstriction via V_2 receptors

35.3 Which of the following statements is **not** true?

(a) the control of body Na^+ content by the kidney is the main regulator of body fluid volume

(b) stretch receptors on the low pressure side of the heart detect changes in the central venous pressure which reflect changes in blood volume

(c) increased pressure in the renal afferent arterioles stimulates renin release

(d) angiotensin II is the primary hormone for Na^+ homeostasis

35.4 Atrial natriuretic peptide (ANP)

(a) is released from left atrial muscle in response to stretching caused by increased arterial blood pressure

(b) suppresses the production of renin and aldosterone but not ADH

(c) inhibits the effects of ADH in the distal nephron and causes renal vasodilatation

(d) causes increased excretion of Na^+ but not water

Chapter 36: Control of acid–base status

36.1 The Henderson–Hasselbalch equation

(a) allows one to determine the molecular weight of a weak acid from its pH alone

(b) is equally useful with solutions of acetic and hydrochloric acid

(c) employs the same value for pK for all weak acids

(d) relates the pH of a solution to the pK and the concentrations of weak acid and conjugate base

36.2 Which of the following statements about buffers is true?

(a) the pH of a buffered solution remains constant no matter how much acid or base is added to the solution

(b) the strongest buffers are those composed of strong acids and strong bases

(c) the maximal buffering capacity occurs when pH = pK

(d) the pH of blood is not maintained by a buffering system

36.3 In the proximal renal tubule

(a) bicarbonate is freely filtered

(b) over 95% of bicarbonate is reabsorbed

(c) bicarbonate is transported into the interstitium largely by Na^+–HCO_3^- antiporters

(d) there is a net absorption of H^+

36.4 Which of the following statements is **not** true?

(a) in the proximal nephron, H^+ secretion promotes HCO_3^- reabsorption

(b) a fall in blood pH will stimulate renal H^+ secretion

(c) renal mechanisms are fast because their capacity to handle H^+ and HCO_3^- is greater than that of the lungs for handling CO_2

(d) hypokalaemia is associated with an enhanced H^+ secretion

Chapter 37: Gastrointestinal tract: overview and the mouth

37.1 The gastrointestinal tract

(a) is made up of the mouth, oesophagus, stomach, pancreas, liver, and small and large intestine

(b) general structure comprises the mucosal layer, the lamina propria, the muscularis mucosa, the submucosal layer, followed by a layer of longitudinal muscle and then a layer of circular muscle

(c) has two layers of smooth muscle between which lies a nerve plexus called Meissner's plexus

(d) has two nerve plexuses called the submucosal plexus (Meissner's plexus) and the myenteric plexus (Auerbach's plexus)

37.2 Mastication

(a) is necessary for most foods to be fully absorbed by the rest of the gastrointestinal tract

(b) usually develops forces of between 1500 and 2000 N between the teeth

(c) can lead to stimulation of periodontal and mucosal mechanoreceptors that in turn lead to the reflex secretion of saliva from the submandibular/sublingual salivary glands but not from the parotid glands

(d) involves the coordinated activity of the teeth, jaw muscles, temporomandibular joint and tongue, as well as the lips, palate and salivary glands

37.3 Which of the following statements is **not** true? Saliva

(a) normally has a very low pH to enhance breakdown of food in the mouth

(b) aids swallowing

(c) contains both inorganic and organic constituents

(d) is hypotonic to plasma when secreted at a low rate

37.4 Swallowing

    (a) involves both voluntary and reflex responses, initiated by the stimulation of mechanoreceptors with afferents in the glossopharyngeal (IX) and trigeminal (V) nerves

    (b) is coordinated by a group of neurones called the swallowing centre found in the thalamus

    (c) occurs after a normal expiration before inspiration begins again

    (d) involves a complex series of events in which the soft palate elevates to prevent food entering the nasal cavity, the larynx is raised and the glottis is closed, and the food pushes the tip of the epiglottis over the tracheal opening to prevent food entering the trachea

Chapter 38: Oesophagus and stomach

38.1 Gastric secretion is

    (a) increased by anticipation of a meal

    (b) increased by acid in the duodenum

    (c) decreased by gastrin

    (d) increased by secretin

38.2 With regard to gastric secretion

    (a) the cephalic phase is mainly brought about by the thought of food

    (b) the gastric phase is mainly stimulated by activity of the vagus nerve

    (c) mechanoreceptors in the stomach wall can be stretched and set up local myenteric reflexes and longer vagal reflexes which stimulate the release of gastrin, histamine, acid, enzymes and mucus

    (d) a high pH in the stomach inhibits gastrin secretion

38.3 Which of the following statements is **not** true?

    (a) the gastric phase of secretion normally lasts for about 3 hours

    (b) the chyme enters the duodenum through the pyloric sphincter

    (c) the presence of chyme in the pyloric antrum causes opening of the pyloric sphincter

    (d) the rate at which the stomach empties depends on the volume in the antrum and the fall in pH of the chyme, both leading to a decrease in emptying

38.4 Which of the following is **not** a function of the stomach?

    (a) to store food temporarily

    (b) to digest food using acids, enzymes and peristaltic movements

    (c) to regulate the release of chyme into the small intestine

    (d) to secrete intrinsic factor which is essential for the absorption of fat-soluble vitamins

Chapter 39: Small intestine

39.1 The parts of the small intestine (start with the part closest to the stomach) are

    (a) duodenum, jejunum, ileum

    (b) duodenum, ileum, jejunum

    (c) ileum, jejunum, duodenum

    (d) ileum, duodenum, jejunum

39.2 The intestinal phase of gastric juice secretion

    (a) is brought about by both distension of the duodenum and the release of secretin

    (b) is probably due to activation of G-cells in the intestinal mucosa

    (c) is activated by the release of secretin and cholecystokinin

    (d) controls gastric emptying

39.3 Which of the following is **not** secreted by the intestinal epithelial cells (brush border)?

    (a) trypsin

    (b) maltases

    (c) sucrase

    (d) lactase

39.4 All of the following are required for fat digestion **except**

    (a) bile acids

    (b) pancreatic lipase

    (c) cholic acid

    (d) monoglycerides

Chapter 40: The exocrine pancreas, liver and gallbladder

40.1 With regard to pancreatic secretion

    (a) pancreatic acinar cells contain trypsin

    (b) when acid enters the duodenum it stimulates pancreatic secretion

    (c) pancreatic secretion is inhibited by gastrin secreted by the G-cells of the antrum

    (d) cholecystokinin inhibits secretion from the exocrine pancreas

40.2 Which of the following statements is true?

    (a) bile is diluted in the gallbladder

    (b) bile salts are mostly absorbed in the distal ileum

    (c) bile salts are formed from the breakdown products of haemoglobin

    (d) the gallbladder can produce 500–1000 mL of bile per day

40.3 The liver

    (a) consists of four lobules

    (b) is involved in metabolism of a vast number of substances produced by the digestion and absorption of food as well as having an important endocrine function

    (c) contains a single column of cells called hepatocytes which secrete hepatic bile which is hypotonic with plasma

    (d) hepatocytes secrete hepatic bile which contains bile salts, bile pigments, cholesterol, lecithin and mucus

40.4 Cholecystokinin

    (a) is a glycoprotein

    (b) is secreted mainly by the distal ileum

    (c) stimulates relaxation of the gallbladder

    (d) is released into the bloodstream and stimulates the secretion of pancreatic enzymes

Chapter 41: Large intestine

41.1 The parts of the large intestine (starting from the part closest to the small intestine) are the

    (a) ascending, transverse, descending and sigmoid colon, caecum, rectum and anal canal

    (b) caecum, descending, transverse, ascending and sigmoid colon, rectum and anal canal

    (c) caecum, ascending, transverse, descending and sigmoid colon, anal canal and rectum

    (d) caecum, ascending, transverse, descending and sigmoid colon, rectum and anal canal

41.2 In the large intestine
   (a) the caecum, and ascending and descending colon are innervated by parasympathetic branches of the vagus nerve
   (b) the descending and sigmoid colon, rectum and anal canal are innervated by the parasympathetic branches of the vagus nerve
   (c) the external anal sphincter is composed of striated muscle and innervated by the pudendal nerve
   (d) stimulation of the sympathetic nerve fibres causes segmental contraction

41.3 The main function of the large intestine is
   (a) to store the residues of food and absorb water and electrolytes
   (b) mixing and propulsion of chyme at a rate of 50–100 cm/h
   (c) the further digestion and absorption of nutrients
   (d) defecation

41.4 The bacteria in the gastrointestinal tract are
   (a) evenly distributed throughout the small and large intestines
   (b) found mainly in the large intestine because of the acid environment
   (c) 99% anaerobic and most are lost in the faeces
   (d) involved in the synthesis of vitamins B_{12} and C

Chapter 42: Endocrine control

42.1 Which of the following is **not** a hormone?
   (a) G-proteins
   (b) insulin
   (c) cortisol
   (d) adrenaline (epinephrine)

42.2 With regard to endocrine control
   (a) the endocrine system provides a rapid, precise but short-term control of cellular function
   (b) paracrine cells release their hormones into the bloodstream so that they reach all parts of the body
   (c) autocrine cells release hormones that control their own function
   (d) hormones are secreted from either classical glandular tissues or whole organs and not from other tissues

42.3 Hormones
   (a) are activated by metabolic transformation by enzymes in the liver or at the site of action
   (b) rely heavily on a positive feedback mechanism for their control
   (c) are involved in homeostasis, reproduction, growth and development and metabolism
   (d) can exert their effect directly on the cell by interaction with non-specific receptor proteins

42.4 Antidiuretic hormone (ADH) is secreted from the ____ and its main target tissue is the ____
   (a) anterior pituitary, kidney
   (b) intermediate pituitary, hypothalamus
   (c) posterior pituitary, kidney
   (d) adrenal cortex, liver and kidney

Chapter 43: Control of metabolic fuels

43.1 In an anabolic state
   (a) storage molecules are broken down
   (b) there is an increase in glucagon and a decrease of insulin

   (c) there is an uptake of glucose and fatty acids by the liver, muscle and adipose tissue
   (d) leptin is released from adipose tissue

43.2 The pancreatic cells that secrete somatostatin are the
   (a) F-cells
   (b) A (or α) cells
   (c) B (or β) cells
   (d) D (or δ) cells

43.3 The effects of insulin on target cells
   (a) are mediated by voltage-gated Ca^{2+} channels
   (b) are to stimulate glycogen uptake
   (c) are to stimulate the release of glucagons
   (d) are mediated by a receptor tyrosine kinase

43.4 The effects of hypoglycaemia do **not** include
   (a) weakness
   (b) blurred vision
   (c) headache
   (d) polyuria

Chapter 44: The hypothalamus and the pituitary gland

44.1 Which hypophyseal structure receives signals from the hypothalamus via the hypophyseal portal vessels?
   (a) neurohypophysis
   (b) supraoptic nucleus
   (c) adenohypophysis
   (d) paraventricular nucleus

44.2 Which of the following hormones is **not** produced by the anterior pituitary?
   (a) follicle-stimulating hormone
   (b) thyroid-stimulating hormone
   (c) oxytocin
   (d) adrenocorticotrophic hormone

44.3 Hypophysiotropic hormones
   (a) originate from the magnocellular neurones in the hypothalamus
   (b) are peptides or proteins released by nerve endings into the bloodstream at the inferior hypophyseal artery
   (c) are not involved in neuroendocrine reflexes
   (d) originate from parvocellular neurones whose cell bodies are found in the anterior pituitary

44.4 Which of the following statements is **not** true?
   (a) hormones released from the hypothalamus tend to appear in the blood in discrete pulses and not as a continuous secretion
   (b) the hypothalamus surrounds the second ventricle at the base of the medial forebrain
   (c) the hypothalamic subdivisions important for endocrine function include the paraventricular, periventricular, supraoptic and arcuate nuclei.
   (d) the release of ADH is controlled by conventional negative feedback mechanisms based on osmolality whereas oxytocin involves positive feedback mechanisms

Chapter 45: Thyroid hormones and metabolic rate

45.1 Concerning thyroid hormones
   (a) T_3 is some 10 times more potent in the tissues than T_4
   (b) T_4 and T_3 are mainly transported in the blood bound to albumin

(c) T_3 is converted into T_4 in the tissues

(d) T_4 and T_3 are secreted by the thyroid in equal quantities

45.2 The primary effect of T_3 and T_4 is to

(a) decrease blood glucose

(b) promote the release of calcitonin

(c) promote heat generating metabolic reactions

(d) stimulate the uptake of iodine by the thryroid

45.3 What hormone would best fit this description? 'affects metabolism of cells; necessary for the CNS to develop properly; necessary for normal bone growth; stored extracellularly'

(a) cortisol

(b) growth hormone

(c) T_4/T_3

(d) thyrocalcitonin

45.4 In mild to moderate cases of hypothyroidism the following symptom is not usually seen

(a) lethargy

(b) sluggishness

(c) intolerance to cold

(d) tissue bloating

Chapter 46: Growth factors

46.1 Growth factors

(a) stimulate mitosis, growth and apoptosis

(b) are modified amino acids

(c) are peptide hormones

(d) are not involved in the stimulation of red blood cell production

46.2 With regard to insulin and insulin-like growth factors (IGF-1 and IGF-2)

(a) they both have a similar function

(b) the IGFs are mitogenic, trophic and act as survival factors

(c) insulin promotes catabolic activity generally

(d) they have quite different structures

46.3 With the exception of ____, growth factors work by activating receptor tyrosine kinases

(a) TGFα, TGFβ and EGF

(b) TGFβ, erythropoietin and EGF

(c) TGFα, erythropoietin and the cytokines

(d) TGFβ, erythropoietin and the cytokines

46.4 The following growth factor has been implicated in the maintenance of colorectal and breast cancers

(a) EGF

(b) FGF-1–24

(c) PDGF

(d) NGF

Chapter 47: Somatic and skeletal growth

47.1 With regard to growth hormone (somatotrophin), which of the following statements is **not** true?

(a) growth hormone provides the main drive for growth spurts during development

(b) growth hormone is a protein released from the anterior pituitary somatotrophs under the control of the hypothalamus

(c) the release of growth hormone varies throughout the day, with the peak following a large meal

(d) growth hormone release is increased in response to exercise and stress

47.2 The overproduction of growth hormone in adults leads to

(a) gigantism

(b) dwarfism

(c) acromegaly

(d) osteoporosis

47.3 The type of bone that has a dense structure and provides most of the strength of the skeleton is called

(a) trabecular bone

(b) cortical bone

(c) diaphyseal bone

(d) epiphyseal bone

47.4 The large cells present in mature bone that remove old bone matrix so that it can be replaced by new material are called

(a) osteoblasts

(b) osteocytes

(c) osteoclasts

(d) macrophages

Chapter 48: Control of plasma calcium

48.1 Which of the following is **not** a hormone involved in the control of Ca^{2+} levels in plasma?

(a) parathyroid hormone

(b) calcitonin

(c) ergocalciferol

(d) 1,25-dihydroxycholecalciferol

48.2 In the regulation of Ca^{2+} balance

(a) the parathyroid glands produce both parathyroid hormone and calcitonin

(b) ionized plasma Ca^{2+} controls parathyroid hormone release

(c) parathyroid hormone has a direct effect on bone causing bone reabsorption

(d) parathyroid hormone is a 64-amino acid peptide and is the major controller of free calcium in the body

48.3 Which of the following statements is **not** true?

(a) vitamin D is an umbrella term for two molecules, ergocalciferol and cholecalciferol

(b) the D vitamins are derived from dehydrocholesterol

(c) D vitamins are converted into 1,25-dihydroxycholecalciferol in the liver

(d) the major action of 1,25-dihydroxycholecalciferol is to enable Ca^{2+} absorption from the gut

48.4 The following hormone(s) can promote the incorporation of calcium into bones

(a) growth hormone

(b) thyroid hormone

(c) sex steroids

(d) all of the above

Chapter 49: The adrenal glands and stress

49.1 Regarding the adrenal gland

(a) the adrenal cortex produces steroids and catecholamines

(b) the zona glomerulosa produces aldosterone

(c) adrenal androgen production is controlled by gonadotrophins

(d) cortisol production is controlled by angiotensin

49.2 Which of the following statements is **untrue**?

(a) the adrenal medulla is functionally part of the sympathetic nervous system

(b) Cushing's syndrome is due to overproduction of glucocorticoids

(c) the adrenal medulla and the adrenal cortex function independently from each other

(d) the chromaffin cells of the adrenal medulla manufacture and secrete adrenaline (epinephrine) and noradrenaline (norepinephrine)

49.3 Cortisol

(a) is produced in the zona glomerulosa

(b) its analogues are collectively classified as mineralocorticoids although they do have glucocorticoid actions

(c) release is stimulated by adrenocorticotropic hormone (ACTH)

(d) is not released during normal physiological activity but in response to stress

49.4 Stress

(a) response is driven by the amygdala which is part of the midbrain

(b) is the primary stimulus in the increased release of mineralocorticoids

(c) can cause increased activity in the thalamus

(d) can cause increased activity in the parasympathetic nerves that leads to acid secretion in the stomach

Chapter 50: Endocrine control of reproduction

50.1 Regarding the control of the reproductive system

(a) gonadotrophin-releasing hormone is secreted by the pituitary

(b) the gonads secrete glucocorticoids

(c) testosterone is a steroid

(d) progesterone is secreted by the ovarian follicle

50.2 Which of the following statements is true?

(a) in the male, luteinizing hormone acts on the Sertoli cells to stimulate the production of testosterone

(b) plasma luteinizing hormone levels are relatively constant minute to minute

(c) the hypothalamus secretes luteinizing hormone

(d) in the female, luteinizing hormone acts on the theca interna cells to stimulate the production of testosterone

50.3 Leydig cells produce

(a) mucus

(b) androgen-binding proteins

(c) testosterone

(d) FSH and LH

50.4 During ovulation all of the following occur **except**

(a) formation of the corpus luteum

(b) oestrogen production is very low

(c) high production of follicle-stimulating hormone

(d) high production of luteinizing hormone

Chapter 51: Sexual differentiation and function

51.1 All of the following are involved in male sexual differentiation during early development **except**

(a) testosterone

(b) progesterone

(c) dihydrotestosterone

(d) *Sry* gene on the Y chromosome

51.2 With regard to puberty

(a) puberty begins with the increased release of luteinizing hormone

(b) the onset of menstruation in females occurs at roughly 47 kg irrespective of age

(c) in males, luteinizing hormone stimulates the release of testosterone from the Sertoli cells

(d) secondary sexual characteristics are stimulated by testosterone

51.3 Which of the following statements is **not** true?

(a) the penis becomes erect as a result of dilation of blood vessels entering the corpora cavernosum and corpus spongiosum

(b) in the penis, the parasympathetic nerves cause vasodilatation by releasing acetylcholine, vasoactive intestinal peptide and nitric oxide

(c) nitric oxide increases the manufacture of cyclic guanosine monophosphate (cGMP) in blood vessel smooth muscle cells to cause them to contract

(d) sildenafil (Viagra) inhibits the breakdown of cGMP and thus enhances erectile function

51.4 An orgasm

(a) is due to a reflex activation of the parasympathetic nerves in both males and females

(b) in males involves the mixing of the sperm and the secretions of the bulbourethral gland, seminal vesicle and prostate gland before they enter the urethra

(c) in males involves the secretion of the sugar glucose to provide energy

(d) in females may involve the release of pituitary oxytocin elicited by mechanical stimulation of the cervix

Chapter 52: Fertilization, pregnancy and parturition

52.1 The unfertilized ovum can survive for____ after ovulation

(a) 1–2 h

(b) 10–12 h

(c) 24 h

(d) 2–3 days

52.2 During fertilization

(a) acid in the uterus triggers capacitation of sperm

(b) the glycoprotein ZP2 acts as a sperm receptor on the zona pellucida

(c) the ovum undergoes electrical depolarization and discharges granules that impair further binding of sperm

(d) after about 10 h the male and female pronuclei fuse

52.3 Human chorionic gonadotrophin (hCG)

(a) begins to be produced about 4 weeks after fertilization

(b) is produced by the corpus luteum after ovulation

(c) maintains the corpus luteum for a short period

(d) signals that a new menstrual cycle should begin

52.4 Release of the hormone relaxin

(a) inhibits the development of additional ova

(b) inhibits uterine contraction

(c) stimulates the development of the endometrium

(d) softens the cervix prior to parturition

Chapter 53: Lactation

53.1 Regarding the control of lactation
  (a) prolactin stimulates milk ejection
  (b) placental steroid secretion inhibits lactation during pregnancy
  (c) vasopressin (ADH) is secreted in response to suckling
  (d) oxytocin secretion can be suppressed by dopamine agonists

53.2 Prolactin
  (a) inhibits release of follicle-stimulating hormone from the pituitary
  (b) levels in the plasma rise following birth and lead to milk production
  (c) is only released in and around birth
  (d) release is maintained after birth mainly by suckling

53.3 Colostrum
  (a) contains high levels of antibodies that provide the infant with basic immunological protection for the rest of its life
  (b) is particularly rich in proteins
  (c) has a higher sugar content than milk
  (d) has a higher fat content than milk

53.4 The main hormone involved in the milk let down reflex is
  (a) progesterone
  (b) oestrogens
  (c) oxytocin
  (d) prolactin

Chapter 54: Introduction to sensory systems

54.1 Transduction describes
  (a) the transmission of signals across synapses by neurotransmitters
  (b) the transmission of nerve impulses from receptors to cortex
  (c) the translation of the stimulus into a code of action potentials
  (d) the generation of conscious mental states from nerve impulses

54.2 The modality of a sensory stimulus is encoded in the somatosensory system by
  (a) the type of receptor activated
  (b) the frequency of action potentials evoked in the afferent nerve fibre
  (c) the rate of receptor adaptation
  (d) the method used to transduce the stimulus at the receptor

54.3 A receptor potential is an example of
  (a) a frequency-coded signal
  (b) a spatially-coded signal
  (c) a temporally-coded signal
  (d) an amplitude-coded signal

54.4 Which of the following statements is **not** true?
  (a) lateral inhibition is normally encountered under pathological conditions (e.g. strychnine poisoning)
  (b) inhibition can prevent the spread of excitation at progressively higher levels of the CNS by a phenomenon called lateral inhibition
  (c) lateral inhibition involves the excitation of inhibitory interneurones
  (d) in almost all sensory systems, higher centres can exert inhibitory effects on all lower levels of the CNS

Chapter 55: Sensory receptors

55.1 Which of the following is not a sensory receptor found in mammals?
  (a) thermoreceptor
  (b) pain receptor
  (c) mechanoreceptor
  (d) chemoreceptor

55.2 A Ruffini-type receptor
  (a) is a rapidly adapting mechanoreceptor found in the non-hairy skin which responds to stretching of the skin
  (b) is a slowly adapting mechanoreceptor found in non-hairy skin which is an intensity detector
  (c) is a slowly adapting mechanoreceptor found in hairy skin which responds to stretching of the skin
  (d) is a moderately rapidly adapting mechanoreceptor found in the hairy skin which responds to touch and pressure

55.3 Thermoreceptors
  (a) respond to both warmth and cold
  (b) respond to either warmth or cold
  (c) are specialized nerve endings with corpuscular structures
  (d) have a large receptive field (over $5\,mm^2$)

55.4 The conscious perception of pain
  (a) is associated with the activation of modality-specific receptors
  (b) may be delayed, despite the seriousness of the trauma
  (c) may elicit autonomic reflexes of nausea, heavy sweating and lowered blood pressure
  (d) all of the above

Chapter 56: Special senses: Taste and smell

56.1 Which tongue papillae are taste buds **not** associated with?
  (a) fungiform
  (b) folliate
  (c) circumvallate
  (d) filiform

56.2 The gustatory receptor cells in taste buds
  (a) have a lifespan of about 28 days
  (b) are neuroepithelial cells arranged in a compact pear-shaped structure
  (c) are all innervated by the same nerve fibre which branches to innervate all the cells in the taste bud
  (d) are innervated by either the glossopharyngeal (IX) nerve or the trigeminal (V) nerve

56.3 The two tissue layers that make up the olfactory organ are the
  (a) supporting cells and basal cells
  (b) olfactory epithelium and lamina propria
  (c) olfactory glands and olfactory receptors
  (d) olfactory epithelium and olfactory receptors

56.4 Humans can distinguish
  (a) over 10 000 different odours
  (b) only about 100 different odours
  (c) about 1000 different odours
  (d) up to 5000 different odours

Chapter 57: Special senses: Vision

57.1 In comparison to the cones, the rods are more
  (a) concentrated in the fovea region
  (b) sensitive to dim light

(c) important for colour vision

(d) sensitive to detail

57.2 Which of the following cells transmit impulses to the rest of the central nervous system via axons in the optic nerve?

(a) ganglion cells

(b) bipolar cells

(c) amacrine cells

(d) horizontal cells

57.3 Ganglion cells

(a) are normally silent unless stimulated by light

(b) respond to a stationary light stimulus rather than to changes of light intensity

(c) can be divided into 50% ON-centre cells and 50% OFF-centre cells

(d) are further subdivided into P and M cells, where P cells are sensitive to contrast and movement and M cells are colour selective

57.4 The axons of the ganglion cells from the_____ region of the retina of the left eye and the _____ region of the retina of the right eye proceed into the left optic tract

(a) temporal, temporal

(b) nasal, nasal

(c) nasal, temporal

(d) temporal, nasal

Chapter 58: Special senses: Hearing and balance

58.1 A young, healthy human can detect sound wave frequencies of between

(a) 10 Hz and 100 kHz

(b) 40 Hz and 50 kHz

(c) 40 Hz and 20 kHz

(d) 5 Hz and 20 kHz

58.2 The order in which sound travels through the auditory system is

(a) external auditory meatus, tympanic membrane, round window, scala tympani, scala vestibuli, oval window

(b) external auditory meatus, tympanic membrane, ossicles, oval window, scala tympani, scala media, round window

(c) external auditory meatus, tympanic membrane, ossicles, round window, scala vestibuli, scala tympani, oval window

(d) external auditory meatus, tympanic membrane, ossicles, oval window, scala vestibuli, scala tympani, round window

58.3 The vestibular system is **not** involved in

(a) balance

(b) hearing

(c) postural reflexes

(d) eye movements

58.4 The utricle send signals representing _____ movements and the saccule conveys information about _____ movements

(a) vertical, forward and backward

(b) forward and backward, vertical

(c) rotational, horizontal

(d) horizontal, rotational

Chapter 59: Motor control and the cerebellum

59.1 The idea of a voluntary movement is thought to originate in the

(a) primary sensory cortex

(b) primary motor cortex

(c) association cortex

(d) cerebellum

59.2 The motor cortex activates the α- and fusi-motor neurones via the

(a) lateral corticospinal tract

(b) corticorubrospinal tract

(c) neither a nor b

(d) both a and b

59.3 Which of the following is **not** a functional and anatomical structure of the cerebellum?

(a) spinocerebellum

(b) cerebrospinocerebellum

(c) cerebrocerebellum

(d) vestibulocerebellum

59.4 The function of the cerebellum is

(a) as a comparator, comparing sensory and motor inputs

(b) as a timing device, converting motor signals into a sequence of events

(c) as a store of motor information

(d) all of the above

Chapter 60: Proprioception and reflexes

60.1 Which of the following mechanoreceptors is **not** a proprioceptor?

(a) Golgi tendon organ

(b) muscle spindle

(c) Ruffini ending in joint

(d) Pacinian corpuscle

60.2 Immediately following the section of both the dorsal and ventral nerve roots supplying a muscle, an increase in the frequency of discharge of a muscle spindle primary afferent from that muscle will be produced by

(a) stimulation of the fusimotor fibres to the muscle distal to the section

(b) stimulation of the α-motor neurone to the muscle distal to the section

(c) contraction of the homonymous muscle

(d) strong stimulation of the proximal end of the cut dorsal root

60.3 The properties of muscle spindles and Golgi tendon organs make them most likely to be detectors of

(a) position sense

(b) sense of movement

(c) force sensation

(d) all of the above

60.4 What kind of reflexes are Golgi tendon organs involved in?

(a) monosynaptic stretch reflexes

(b) withdrawal reflexes

(c) polysynaptic stretch reflexes

(d) protective reflex

Answers to self-assessment MCQs

1.1 b, 1.2 d, 1.3 c, 1.4 a

2.1 c, 2.2 d, 2.3 b, 2.4 a

3.1 b, 3.2 c, 3.3 a, 3.4 a

4.1 c, 4.2 b, 4.3 d, 4.4 b

5.1 b, 5.2 b, 5.3 a, 5.4 a, 5.5 b

6.1 c, 6.2 d, 6.3 b, 6.4 c

7.1 b, 7.2 c, 7.3 c, 7.4 c

8.1 a, 8.2 d, 8.3 a, 8.4 b

9.1 a, 9.2 b, 9.3 d, 9.4 c

10.1 a, 10.2 c, 10.3 c, 10.4 d

11.1 c, 11.2 a, 11.3 a, 11.4 c

12.1 c, 12.2 c, 12.3 d, 12.4 b

13.1 b, 13.2 b, 13.3 d, 13.4 c

14.1 a, 14.2 c, 14.3 c, 14.4 d

15.1 c, 15.2 d, 15.3 d, 15.4 c

16.1 c, 16.2 d, 16.3 d, 16.4 d

17.1 c, 17.2 a, 17.3 c, 17.4 d

18.1 b, 18.2 d, 18.3 d, 18.4 b

19.1 d, 19.2 b, 19.3 c, 19.4 c

20.1 b, 20.2 b, 20.3 c, 20.4 d

21.1 c, 21.2 c, 21.3 b, 21.4 b

22.1 d, 22.2 b, 22.3 b, 22.4 c

23.1 a, 23.2 d, 23.3 c, 23.4 c

24.1 a, 24.2 c, 24.3 b, 24.4 c

25.1 c, 25.2 a, 25.3 d, 25.4 d

26.1 a, 26.2 a, 26.3 b, 26.4 d

27.1 a, 27.2 c, 27.3 d, 27.4 c

28.1 b, 28.2 a, 28.3 a, 28.4 d

29.1 b, 29.2 a, 29.3 d, 29.4 c

30.1 b, 30.2 b, 30.3 a, 30.4 d

31.1 d, 31.2 d, 31.3 c, 31.4 b

32.1 a, 32.2 b, 32.3 d, 32.4 c

33.1 d, 33.2 c, 33.3 b, 33.4 c

34.1 d, 34.2 c, 34.3 b, 34.4 c

35.1 d, 35.2 c, 35.3 c, 35.4 c

36.1 d, 36.2 c, 36.3 a, 36.4 c

37.1 d, 37.2 d, 37.3 a, 37.4 d

38.1 a, 38.2 c, 38.3 d, 38.4 d

39.1 a, 39.2 b, 39.3 a, 39.4 d

40.1 b, 40.2 b, 40.3 d, 40.4 d

41.1 d, 41.2 c, 41.3 a, 41.4 c

42.1 a, 42.2 c, 42.3 c, 42.4 c

43.1 b, 43.2 d, 43.3 d, 43.4 d

44.1 c, 44.2 c, 44.3 c, 44.4 b

45.1 a, 45.2 c, 45.3 c, 45.4 d

46.1 c, 46.2 b, 46.3 c, 46.4 a

47.1 c, 47.2 c, 47.3 b, 47.4 c

48.1 c, 48.2 b, 48.3 c, 48.4 d

49.1 b, 49.2 c, 49.3 c, 49.4 d

50.1 c, 50.2 d, 50.3 c, 50.4 b

51.1 b, 51.2 a, 51.3 c, 51.4 d

52.1 c, 52.2 c, 52.3 c, 52.4 d

53.1 b, 53.2 d, 53.3 b, 53.4 c

54.1 c, 54.2 a, 54.3 d, 54.4 a

55.1 b, 55.2 c, 55.3 b, 55.4 d

56.1 d, 56.2 b, 56.3 d, 56.4 a

57.1 b, 57.2 a, 57.3 c, 57.4 d

58.1 c, 58.2 d, 58.3 b, 58.4 b

59.1 c, 59.2 d, 59.3 b, 59.4 d

60.1 d, 60.2 a, 60.3 c, 60.4 d

Physiology at a Glance, Third Edition. Jeremy P.T. Ward and Roger W.A. Linden.

148 © 2013 John Wiley & Sons, Ltd. Published 2013 by John Wiley & Sons, Ltd.

Appendix I: Comparison of the properties of skeletal, cardiac and smooth muscle

Characteristic	Skeletal muscle	Cardiac muscle	Smooth muscle
Cell shape and size	Long cylindrical cells up to 30 cm long and 100 μm wide	Irregular, branched, rod-shaped cells up to 100 μm long and 20 μm wide	Spindle-shaped cells up to 400 μm long and 10 μm wide
Nuclei	Multinucleated	Mostly single nuclei	Single nuclei
Presence of actin and myosin filaments	Yes	Yes	Yes
Striated (presence of sarcomeres)	Yes	Yes	No
Myogenic activity	No	Yes	Yes
Initiation of contraction	Extrinsic (somatic, neural)	Intrinsic (muscle origin) but influenced by extrinsic autonomic (sympathetic and parasympathetic)	Can be intrinsic via plexus of nerves or extrinsic via autonomic (sympathetic and/or parasympathetic), hormones or stretch
Basic muscle tone	Neural activity	None	Both intrinsic and extrinsic factors
Speed of contraction	Fast	Slow	Very slow
Type of contraction	Phasic	Rhythmic	Tonic with some phasic
Electronic coupling between cells (gap junctions)	No	Yes	A few in multiunit and many in unitary
Influence of hormones on contraction	Small	Large	Large
Effect of nerve stimulation	Excitatory	Excitatory or inhibitory	Excitatory or inhibitory
Spontaneous electrical activity	No	Yes	Unitary (yes), multiunit (no)
Extent of innervation	Each cell innervated	Variable	Unitary (sparse), multiunit (almost every cell)
Site of calcium regulation of contraction	Troponin thin filament	Troponin thin filament	Myosin thick filament
Mechanism of excitation–contraction coupling	Via action potentials and T-system	Via action potentials and T-system	Via action potentials, calcium channels and/or second messengers
Source of activating calcium	Sarcoplasmic reticulum	Sarcoplasmic reticulum and some extracellular	Sarcoplasmic reticulum and some extracellular

Appendix II: Normal physiological values

Blood fluid volumes

Blood volume	70 mL/kg body weight
Plasma volume	40 mL/kg body weight
Total body fluid volume	60% body weight (males)
	50% body weight (females)
Intracellular fluid volume	65% of total body fluid volume
Extracellular fluid volume	34% of total body fluid volume

Blood cells

Haematocrit	45% (range 40–50) males
	40% (range 36–46) females
Haemoglobin concentration	150 g/L (range 130–170) males
	140 g/L (range 120–150) females
Red cell (erythrocyte) count	5×10^{12}/L (range 4.5–6.5) males
	4.5×10^{12}/L (range 3.8–5.6) females
Reticulocyte count	2% of erythrocyte count
Erythrocyte sedimentation rate (ESR)	<5 mm/h (males)
	<7 mm/h (females)
White cells count	7×10^9/L (range 4–11)
Neutrophils	3.5×10^9/L
Lymphocytes	2×10^9/L
Eosinophils	0.2×10^9/L
Monocytes	0.5×10^9/L
Basophils	$<0.1 \times 10^9$/L
Platelet count	2.5×10^{11}/L (range 1.4–4.0)

Plasma

Plasma protein concentration	60 g/L
Plasma oncotic (colloid osmotic) pressure	25 mmHg
Plasma osmolality	290 mosmol/kg H_2O
Na^+	140 mmol/L
K^+	4 mmol/L
Ca^{2+}	1 mmol/L (free)
Cl^-	108 mmol/L
HCO_3^-	25–30 mmol/L

Cardiovascular function

Cardiac output, rest	5 L/min
Cardiac output, exercise	>20 L/min
Heart rate, rest	70/min
Heart rate, exercise	180/min
Resting stroke volume	70 mL
Systemic arterial blood pressure	110 mmHg (systolic)
	80 mmHg (diastolic)
Pulmonary arterial pressure	25 mmHg (systolic)
	15 mmHg (diastolic)
Central venous pressure	5 cmH$_2$O (range 3–8)
Mean arterial pressure	90 mmHg
Mean blood pressure at start of arterioles	65 mmHg
Capillary pressure	25–35 mmHg (arterial side)
	15 mmHg (venous side)
End diastolic volume (EDV)	130 mL (range 120–140)
End diastolic pressure	<10 mmHg

Physiology at a Glance, Third Edition. Jeremy P.T. Ward and Roger W.A. Linden.

© 2013 John Wiley & Sons, Ltd. Published 2013 by John Wiley & Sons, Ltd.

Percentage of cardiac output to various organs at rest

Brain	14%
Heart	4%
Liver and digestive system	27%
Kidney	20%
Skeletal muscle	21%
Skin	5%
Bone and other tissues	9%

Respiratory function

Static lung compliance	$1.5\,L/kPa$
Intrapleural pressure (quiet breathing)	$-4\,cmH_2O$ (expiration)
	$-9\,cmH_2O$ (inspiration)
Tidal volume	$500\,ml$
Respiratory rate (rest)	$12/min$
Respiratory minute volume	$6\,L/min$
Dead space	$150\,mL$
Alveolar ventilation rate	$5\,L/min$
FEV_1/FVC	80% (range 75–90)
Vital capacity (VC)	$5.5\,L$
Inspiratory reserve volume (IRV)	$3.3\,L$
Expiratory reserve volume (ERV)	$1.7\,L$
Total lung capacity (TLC)	$7.3\,L$
Functional reserve capacity (FRC)	$3.5\,L$
Residual volume (RV)	$1.8\,L$

Blood gases and acid–base balance

Systemic arterial pH	7.36 (range 7.3–7.42)
Systemic arterial Po_2	$13\,kPa$ ($98\,mmHg$)
Systemic arterial Pco_2	$5.3\,kPa$ ($40\,mmHg$)
Base excess	-2 to $+2\,mmol/L$
Mixed venous blood Po_2 (resting)	$5.3\,kPa$ ($40\,mmHg$)
Mixed venous blood Pco_2 (resting)	$6.1\,kPa$ ($46\,mmHg$)
Alveolar Po_2	$13.3\,kPa$ ($100\,mmHg$)
Alveolar Pco_2	$5.3\,kPa$ ($40\,mmHg$)
Capillary Po_2	$5.3\,kPa$ or less ($40\,mmHg$ or less)
Capillary Pco_2	$6.1\,kPa$ or more ($46\,mmHg$ or more)

Renal function

Renal blood flow	$1.2\,L/min$
Renal plasma flow	$600\,mL/min$
Glomerular filtration rate	$125\,mL/min$
Average urine output	$1\,mL/min$
Renal glucose threshold	$11\,mmol/L$

Gastrointestinal function

Fluid inputs into the digestive system (total = 9 L/day)

Food and drink	$2.0\,L$
Saliva	$1.5\,L$
Bile (liver)	$0.5\,L$
Gastric secretions	$2.0\,L$
Pancreatic secretions	$1.5\,L$
Intestinal secretions	$1.5\,L$

Fluid removed from the digestive system (total = 9 L/day)

Absorption from small intestine	$7.5\,L$
Absorption from large intestine	$1.4\,L$
Excreted in faeces	$0.1\,L$

(Continued)

Nervous system

Equilibrium potential for Na^+ — $+65\,mV$
Equilibrium potential for K^+ — $-90\,mV$
Resting membrane potential in excitable cells — $-70\,mV$ (range -60 to -90)
Resting membrane potential in non-excitable cells — $-10\,mV$

Blood glucose levels

Normal

Fasting pre-prandial — 4.0–5.9 mmol/L
Post-prandial (2 h after meal) — <7.8 mmol/L

Diabetic fasting pre-prandial

Pre-diabetes or impaired glucose glycaemia — 6.0–6.9 mmol/L
Diagnosis of diabetes — >6.9 mmol/L
HbA_{1C} (glycated haemoglobin), a measure of the average plasma glucose concentration
Normal — 20–41 mmol/L (4–5.9%)
Diabetic patients — >47 mmol/L (>6.5%)

Index

Note: page numbers in *italics* refer to figures and tables.

Physiology at a Glance, Third Edition. Jeremy P.T. Ward and Roger W.A. Linden.
© 2013 John Wiley & Sons, Ltd. Published 2013 by John Wiley & Sons, Ltd. **153**

stomach *84*, *86*, 87
store operated channels (SOC) *52*, 53
stress
    catecholamines 109
    glucocorticoids 109
    growth hormone (GH) *104*, 105
    hormones 98
    lactation failure 117
    responses *108*, 109
stretch receptors 69, 81, *118*, *130*, 131
striation 35, 41
stroke rate 51
stroke volume (SV) 43, 47, *50*, 51, *150*
stroke work 47
subclavian vein 57
sublingual gland *84*, 85
submandibular gland *84*, 85
submucosal glands *60*, 61
submucosal plexus *84*, 85
substance P 53
suckling *116*, 117
sucrase 89
summation 39
superior colliculus 125
supplementary motor area *128*, 129
surface tension, alveoli 63
surfactant 33, 61, *62*, 63
swallowing *84*, 85, 87
sweat glands 24, 25, *58*, 59
sympathetic chains 24, 25
sympathetic cholinergic neurones 25, 59
sympathetic nervous system
    adrenal glands 25, 109
    functions *24*, 25
    heart rate modulation *50*, 51, 55
    intestines 93
    stress *108*, 109
    thermoregulation *58*, 59
symporters 19, *76*, 77
synapses 25, *118*, 119
synaptic cleft 24, *36*, 37
syncope 51, 55
syncytium, functional 41
synergy 69
systemic circulation 43, 45
systole *42*, 43, *46*, 47
systolic pressure 33, *42*, 43

T cells 27, *30*, 31
T helper cells *30*, 31
T wave *46*, 47
tactile discs 121
taeniae coli *92*, 93
taste 85, 119, *122*, 123
taste buds *122*, 123
taste pore *122*, 123
tear glands 24
tectospinal tract 129
teeth 85
temporal coding 119
temporomandibular joint (TMJ) 85
tenase 29
tendons *34*, 35, *130*, 131
TENS (transcutaneous electrical nerve stimulation) 121
terminal arterioles *56*, 57
terminal cisternae *34*, 35, 49
testes 94, *110*, 111, *112*

testosterone
    binding proteins 95
    conversion 111
    fetal development 113
    Leydig cells *110*, 111
    menstrual cycle *110*, 111
    puberty 113
    secretion *94*
    structure *110*
    targets *94*
tetanus *38*, 39
tetany, hypocalcaemic 107
thalamus
    motor control *128*
    sensory pathways 119
    vision 125
thalassaemia 27
Thebesian veins 45
theca interna cells *110*, 111
thermogenesis *100*, 101
thermoreception 123
thermoreceptors *58*, 59, *100*, 121
thermoregulation 43, *58*, 59, 121
thiamine 93
thirst *80*, 81, *97*
thoracic cage 61
thoracic duct 57, 89
thoroughfare vessels *56*, 57, 59
threshold potential
    muscles *36*
    sensory systems *118*, 119
    value 19, *20*, 21
    ventricular myocytes *48*, 49
thrombin *28*, 29
thrombomodulin 29
thromboplastin *see* Tissue Factor
thrombosis 29, 53
thromboxane A (TXA$_2$) *28*, 29, *52*, 53, 59
thymus 31
thyroglobulin *100*, 101
thyroid gland *94*, *100*, 101, *106*, 107
thyroid hormone receptors 101
thyroid hormones 95, *106*, 107
thyroid-responsive element (TRE) 101
thyroid-stimulating hormone (TSH) *94*, 98, *100*, 101
thyroperoxidase 101
thyrotrophin-releasing hormone *98*
thyroxine (T4) *94*, *100*, 101
tidal volume (TV) *60*, 61
tight junctions
    capillaries 57
    kidney tubules *72*, 73, 77
    large intestine *92*, 93
    taste buds *122*
tissue factor (TF, thromboplastin) *28*, 29
tissue plasminogen activator (tPA) 29
tongue *84*, 85, *122*, 123
tonic activity 41
tonic depression 55
total lung capacity (TLC) *60*, 61
total peripheral resistance (TPR) *50*, 51, *54*, 55, 59
trabeculae 45
trachea *60*, 61, *84*, 85
transcapillary exchange 57
transcellular fluid 15
transcription, genetic 17, 101, *102*, 103, 105, 107
transduction *104*, 105, 117, 121, *122*, 123

transfer factor 65
transferrin *26*, 27
transforming growth factor α (TGF-α) 103
transforming growth factor β (TGF-β) *102*, 103, 105
translation, genetic 13
transmembrane proteins 16, 17, *18*, 19
transmural pressure 33, *62*, 63
transport *see also* diffusion; flow; gas exchange
    carbon dioxide 27
    iron *26*, 27
    oxygen 27, *42*, 43
    primary/secondary active *18*, 19, *76*, 77
transporter proteins 17, *18*, 19, 27
transudation 63
Treppe effect 49
tricuspid valve *44*, 45, 47
trigeminal nerve *122*, 123
triglycerides *see* fats
tri-iodothyronine (T3) *94*, *100*, 101
trophoblasts *114*, 115
tropomyosin *34*, 35, 41
troponin *34*, 35, 51
trypsin 89, 91
T-tubules *34*, 35, 37, *40*, 48, 49
tubular transport maximum 77
tunica adventitia *52*, 53
tunica intima *52*, 53
tunica media *52*, 53
turbinate bones 123
turbulent flow *32*, 33
tympanic membrane *126*, 127
tyramine 25
tyrosine *108*
tyrosine kinases 95, *96*, 97, 103

ultrafiltration 73, 75
uncoupling proteins (UCPs) 101
uniporters 19, 77, 79
uptake-1 and -2 25
urea *76*, 77, *78*, 79
ureter *72*, 73
urethra 73, *112*, 113
urethral sphincter 73
urine *see also* excretion
    clearance 75
    concentration *78*, 79
    diabetes insipidus 81
    glucose 81
    hydrogen ions (H$^+$) 83
    output 73, 81
    phosphate (PO$_4^{3-}$) 83
    polyuria 97
uterus
    autonomic nervous system (ANS) *24*
    hormones *94*, *110*, 111
    orgasm 113
    pregnancy *114*, 115
    puberty *112*, 113
    structure *112*
utricle 127

v waves *46*, 47
vagina *112*, 113
vagus nerve
    aortic bodies *68*, 69
    bile 91
    blood pressure regulation *54*, 55

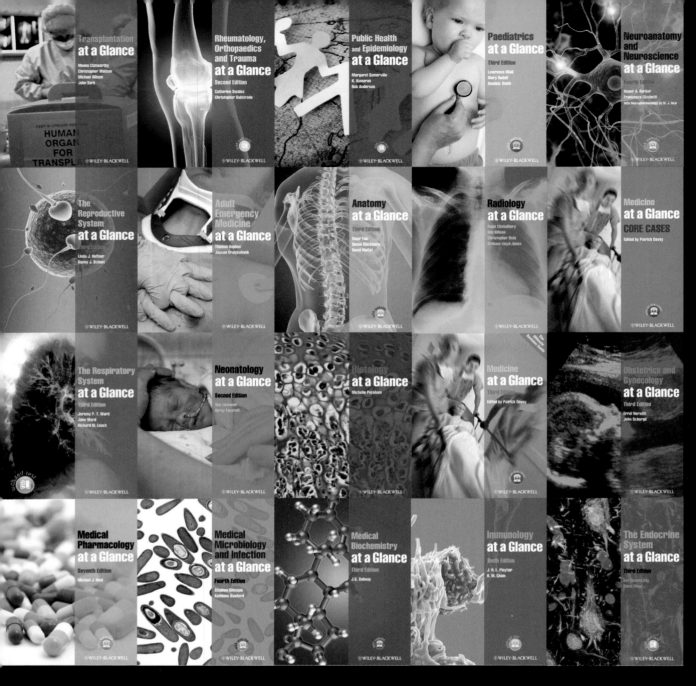

The at a Glance series

Popular double-page spread format • Coverage of core knowledge
Full-colour throughout • Self-assessment to test your knowledge • Expert authors

www.wileymedicaleducation.com

WILEY-
BLACKWELL